GEOLOGY UNDERFOOT IN SOUTHERN CALIFORNIA

Robert P. Sharp
and
Allen F. Glazner

Mountain Press Publishing Company
Missoula, Montana
1993

Third Printing, May 1996

Cover illustration © Deb Santo
Cover design by Cari Reichert

Maps by Allen F. Glazner and Trudi Peek

Library of Congress Cataloging-in-Publication Data

Sharp, Robert P. (Robert Phillip)
Geology underfoot in Southern California / Robert P. Sharp and Allen F. Glazner.
p. cm.
Includes index.
ISBN 0-87842-289-7
1. Geology—California, Southern. 2. n-us-ca. I. Glazner, Allen F. II. Title.
QE90.S65S53 1993 93-16941
557.94'9—dc20 CIP

Printed in the U.S.A.

Mountain Press Publishing Company
P.O. Box 2399 • Missoula, MT 59806
406-728-1900 • 800-234-5308

To my paternal grandfather James M. Sharp
who made my college education possible.

—Robert P. Sharp

To my father George Glazner whose help and
support made my field excursions possible.

—Allen F. Glazner

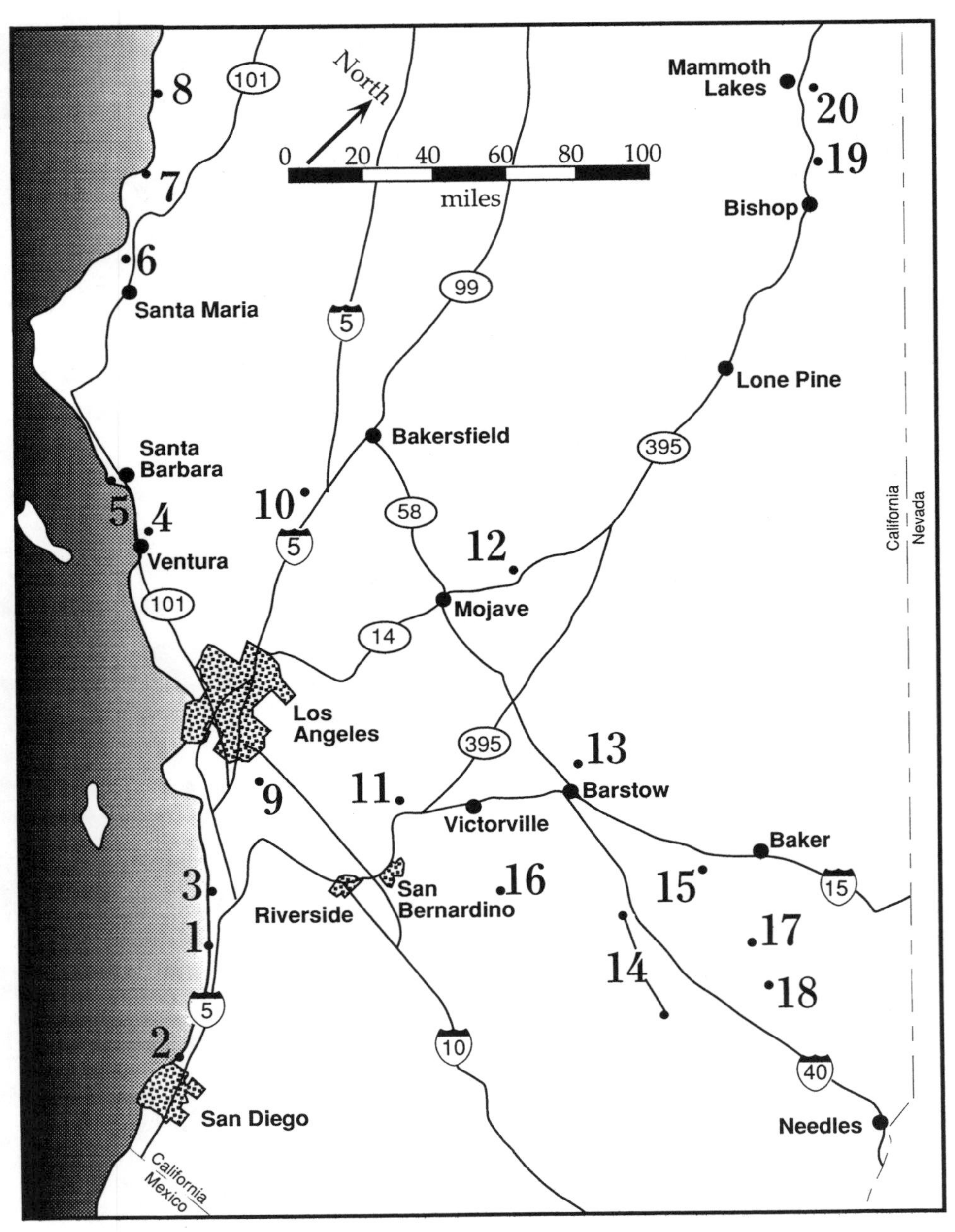

Sites featured in geological vignettes. Map numbers correspond to vignette numbers.

Contents

Preface

A vignette can be an ornamental sketch, a striking picture, a vivid scene, or a short story. Vignettes presented herein are stories, like snapshots, focused on some particular scene, relationship, or feature selected from the rich mixture of southern California's geological phenomena. Like snapshots, they do not cover everything, only items in focus and within the field of view.

Each vignette treats a geological subject of particular interest and significance. Some involve features well known to many readers. Others treat relationships with which readers may be unacquainted or to which they have not given previous thought. Some treat matters of practical concern: San Onofre nuclear plant, Santa Barbara harbor, Ventura oil field. Many offer pure education and enjoyment: Rainbow Basin, Hot Creek, Torrey Pines, San Simeon beach pebbles. A few involve some limited, largely easy, walking; some require careful observation; some can be appreciated from the car.

Topics selected are geographically dispersed but purposely located near heavily populated or frequently visited areas. Some extensively documented areas, such as Yosemite Valley and Death Valley, are not treated. The spectrum of available geological vignettes is so broad and varied that some readers are bound to be disappointed that their favorite spot is not included here. To them we apologize. Selecting just twenty out of a hundred or more outstanding candidates was a traumatic task. Our aim is to present a description and discussion of geological features in simple conversational style, comfortable for non-professional readers. We have tried to keep technical jargon to a minimum. It is difficult to communicate about geological phenomena, however, without some rudimentary knowledge of a few basic principles. Such information follows. Reference to the comprehensive glossary also helps. For each vignette we show the local geography with a map or two. It is helpful to also bring along good local road maps, such as those published by the Automobile Club of Southern California, and topographic maps. To find U. S. Geological Survey topographic maps call map companies listed in the yellow pages. Many maintain a good stock, especially of local maps. We are much in debt to John S. Shelton for generously furnishing nine of his superb aerial photographs, to Mary Olney for her fossil sketches, and to Ed Abbey for providing information about Mitchell Caverns; to Dave Alt and Kathleen Ort for editorial guidance; to Victor Church, Phillip West, Patrick Abbott, Dana Loomis, Bill Wise, and Steve Lipshie for information or comments on specific vignettes, and to Mary Crump, Janice Mayne, Enid Bell, Linda Carson, Nancy Jenkins, and Mary Olney for secretarial, editorial, and drafting services. Material for vignettes has been drawn from many publications and field trips, too numerous to acknowledge individually. The authors of all are saluted with warm thanks. We enjoyed putting this volume together. It is designed to give readers pleasure in becoming acquainted with natural features of the world around us.

THE BIG PICTURE

Rocks tell stories that can be read like pages in a book. Composed of minerals and, in some volcanic rocks, glass, they record the processes that created and continue to shape our modern landscapes—the geology underfoot. Rocks come in three main varieties—igneous, sedimentary, and metamorphic—based on their composition and texture.

Igneous rocks are primary, the original source of other rocks. In fact, all parts of the earth were molten at one time or another early in geologic history. To quote an old professor, who each year began his introductory geology course with this solemn announcement: "In the beginning, all things were igneous."

Igneous rocks form as molten magma cools and crystallizes. Some igneous rocks crystallize below the surface, others after they erupt from volcanoes. Either way, they vary greatly in composition and texture. Granite and rhyolite, for example, have the same chemical composition, but they look vastly different. Granite is an intrusive rock that cools and crystallizes below the surface; it consists of crystals large enough that you can easily see them without a magnifier. Rhyolite is a volcanic rock that cools very quickly; it consists mostly of microscopic crystals, in some cases of completely non-crystalline glass. Granite and rhyolite are both pale rocks, rich in silica and poor in iron and magnesium.

Basalt is another volcanic rock, the most common kind. Like rhyolite, it is a mass of microscopic crystals. Unlike rhyolite, basalt is black. It contains much less silica than granite and rhyolite and a great deal more iron and magnesium.

Sedimentary rocks are secondary. They form through the mechanical breakup and chemical decomposition of other rocks, any kind of other rock. The sediments that eventually solidify into sedimentary rocks are commonly deposited in layers. Consider sandstone. It starts out as layers of sand deposited wherever layers of sand accumulate: on a floodplain, on a beach, in a dune field. In one way and another, the sand grains finally stick together, and sand becomes sandstone.

Metamorphic rocks are also secondary. They form through recrystallization of other rocks, largely without melting. Heat is the prime mover; all metamorphic rocks recrystallize at fairly high temperature, most at something approaching a red heat. Most metamorphic rocks also recrystallize under high pressure, generally while the rock is changing shape. Marble and schist are good examples of metamorphic rocks common in many parts of southern California.

A Look Within

Rocks and the landscapes that contain them make sense only in the light of the internal forces that deform the earth's crust. The earth has a core with a radius of roughly 2,200 miles. Enclosing that is the mantle, about 1,800 miles thick. On the outside is the crust on which we live, at most about 40 miles thick, no thicker in proportion to the size of the earth than the skin on a large apple.

Mantle rocks are very slightly radioactive. They keep themselves warm. In fact, they produce a bit more heat than they can lose through simple conduction. So the rocks in the upper mantle slowly move in great convection cells in which hot material rises, moves laterally as it cools, and then plunges back into the mantle. You can see the same thing happening on a small scale if you watch a cup of coffee stir itself as it cools. Rocks in the mantle are extremely viscous, so they move slowly, a few inches per year. Slow as that movement is, it severely deforms the earth's crust.

The crust and the uppermost part of the mantle, the outer 60 miles or so of the earth, make up the lithosphere. Rocks in that outer zone are colder than those below, and therefore more rigid. Think of the lithosphere as an outer rind on the earth.

The lithosphere is broken into pieces called plates, a dozen or so, depending upon who does the counting. Some of the plates are very large, others quite small. In southern California, we think mostly of the huge Pacific plate, which underlies most of the Pacific Ocean, and the North American plate, which extends from the middle of the Atlantic to the west coast. Both include areas of continent and ocean, passengers on the plates. Oceanic crust is normally about three miles thick, continental crust about 20 miles thick. The rest of the plate, down to a depth of about 60 miles is mantle rock.

Beneath the rigid plates lies the asthenosphere, a weak layer in which the mantle is partially molten. It may be hundreds of miles thick. Plates of the lithosphere definitely move, probably sliding on the weak mantle rocks of the asthenosphere. Different plates travel at different speeds; some poke along at a fraction of an inch per year, others speed as much as four inches per year. Two inches per year is about average.

Each plate moves in its own direction, towards its own destination. Plates move like bumper cars, with no coherent pattern, pulling away from each other, colliding, or sliding past each other.

Where plates pull away from each other, they form an oceanic ridge that you can see on the map of the seafloor—the mid-Atlantic ridge, for example. Fissures open in the crest of the ridge as the plates separate, and molten basalt lava erupts to make new oceanic crust.

Where plates collide, they make a long, trenchlike ocean deep on the map, with a chain of volcanoes 50 to 100 miles away trending parallel to the trench. The Aleutian trench and its associated volcanoes are a good example. Watching a collision between two lithospheric plates is probably as close as we will ever come to seeing the mythical encounter between an irresistible force and an immovable object. The earth solves that dilemma by letting one plate sink into the mantle, invariably one that has oceanic crust on its surface. The oceanic slab sinks because it is denser than the one with continental crust. The ocean deep is where the sinking plate bends down to start its long plunge into the hot depths of the earth. Geologists call these trenches subduction zones. They swallow old oceanic crust at the same rate that eruptions at oceanic ridges create new. Subduction normally causes a lot of igneous activity within the overriding plate.

Where plates slide past each other, they create enormous faults, such as the San Andreas fault, that allow horizontal movement. Geologists call them transform plate boundaries. Although they do not spawn volcanic chains, transform boundaries do cause earthquakes as they jam the rocks on either side into folds and break them along faults. They may also cause some local volcanic activity. The San Andreas fault slices through the southern two-thirds of California, separating the Pacific plate on the west from the North American plate on the east. The Pacific plate is moving north at a rate of about two inches a year. Geologists call the San Andreas a right-lateral fault based on its sense of movement; if you stand on one side of the fault and look across it, the opposite side is moving to your right. In other words, Los Angeles, on the Pacific plate, is approaching San Francisco, on the North American plate, in short jerks to an accompaniment of earthquakes. The San Andreas fault provides Californians with a dynamic and unstable geologic environment—and a stunning natural landscape. It brings them occasional grief and a great deal of pleasure.

We said that plates move by sliding on the weak rocks of the asthenosphere. This may explain how they move, but certainly not why. That question is still open. Years ago, most geologists thought the drag of convection cells in the mantle moved plates. Some call that mantle drag. Now, geologists also consider that plates may glide downslope, away from the crests of oceanic ridges. Some call that ridge push. And lithosphere sinking through an ocean trench because it is denser than the hotter rocks below may drag the rest of the plate along behind. Some call that slab pull. It does seem clear that the fastest plates are those subject to significant ridge push and slab pull and possibly also mantle drag.

Keeping Geologic Time

The question of time hovers over every geologic discussion. Astronomers and geologists, starting from altogether different kinds of evidence, agree that the solar system and the earth are about 4.6 billion years old. Geologists divide that yawning abyss of time into four major subdivisions, called eras. From oldest to youngest they are: Precambrian, which lasted from 4.6 billion to 570 million years ago; Paleozoic, which lasted from 570 to 240 million years ago; Mesozoic, which lasted from 230 to 66 million years ago; and Cenozoic, which began 66 million years ago. In these vignettes, we focus mainly on events of Cenozoic time and the rocks and landscapes they created. However, California contains rocks of all eras, the oldest known being about 1.8 billion years old.

Suppose we let a yardstick represent all of geological time. Precambrian time would measure 30.5 inches, Paleozoic time 3.6 inches, Mesozoic time 1.4 inches, and Cenozoic time about 0.5 inch. The human race originated about 2.5 million years ago, only 0.02 inch down the yardstick. We need to recognize that geologic time is very long, and our time very short. Processes that seem painfully slow to us can work profound changes, given time in geologic measure.

How do we measure geologic time? In many ways, but the simplest and most reliable are geologic "clocks" that run within the rocks as radioactive elements decay into other elements. The proportion of the radioactive parent element to the stable daughter product measures the age of the mineral or rock. To determine the age is a matter of precise chemical analysis. The most useful clocks involve the slow disintegrations of potassium to argon, rubidium to strontium, and uranium to lead.

Earth-shaping Events in Southern California

Southern California is widely regarded as an exotic place in terms of culture, economics, politics, and society. Its geology is also varied, dynamic, even flamboyant. The region is one of the most rapidly deforming areas in North America, if not the world. This tectonic deformation creates a varied landscape and some disturbing earthquakes. One can wonder what role this unusual geology may have played in shaping southern California's freewheeling culture.

The pages of southern California's geologic history book go back about 1.8 billion years, to a time when granites and metamorphic rocks formed the core of an ancient continent. Little is known about this early time because only small remnants of these old rocks remain. The well-displayed record began about 800 million years ago, when a great piece of the ancient continent broke away from what is now North America. Some people think that the other piece now forms part of Antarctica. When the continent split, a new ocean basin formed, and sediments, which later become sedimentary rocks, were laid down along the new continental margin. In places, these rocks accumulated to thicknesses of several miles as the continental margin slowly sank. The Inyo Mountains and the ranges around Las Vegas contain thick layers of sedimentary rocks that date back to this time. The caves of Mitchell Caverns (Vignette 18) are also sculpted in such rocks. The Grand Canyon region preserves a thinner sequence.

This depositional coastline persisted through Paleozoic time. During this period, southern California was in a situation not unlike that off the east coast today—drowned in shallow water and slowly taking on a greater and greater load of sediment. The ancient coastline extended from the Mojave Desert through southern Nevada and on to northern Utah, and fluctuated greatly as sea level rose and fell. Sometimes the sea moved as far inland as the Grand Canyon.

About 240 million years ago, around the beginning of the Mesozoic, the situation changed drastically. A new subduction boundary formed as plates started to dive under the new western edge of North America. The exact cause of this massive plate reorganization is unknown, but it dramatically changed the landscape of California. Subduction typically causes volcanism and in this case generated a north-south chain of vigorous volcanoes along the length of what is now California. The granites of the Sierra Nevada, Peninsular Ranges, and Mojave Desert are the roots of these old volcanoes, and remnants of the volcanoes themselves can be found in places. Much of the crust of present-day California was built by the new volcanism, and California rose from the sea to become largely dry again. Rocks of the Coast Ranges and the Franciscan complex (Vignettes 3 and 8) are sediments that were scraped off as the oceanic plates dove beneath North America.

Construction of the Sierra Nevada and Peninsular Ranges continued until about 85 million years ago, when volcanism stopped, probably because the oceanic plates plunged beneath North America at a shallower angle and didn't generate magma. This is happening today in parts of the Andes.

The period from 85 to about 25 million years ago is a mystery—few rocks of this age exist in inland southern California, so there is little record of what happened. Along the coast, some marine sediments of this age are exposed (Vignette 2). By all accounts, California got a much-needed rest after the Mesozoic magmatic madness. It needed to rest up for what was to come.

A massive blast of volcanism and faulting began about 25 million years ago. The plates reorganized again, subduction stopped, and the San Andreas transform margin formed. This wave of volcanism and faulting moved from south to north at a few inches per year, passing through southernmost California about 25 to 30 million years ago and through Las Vegas about 15 million years ago. North of the wave, normal subduction volcanoes reestablished themselves, and south of it the ever-lengthening San Andreas fault grew. At present, the boundary between subduction and the San Andreas fault is in northern California; Lassen Peak is the southernmost of the active volcanoes.

The wave of faulting and volcanism passed through the Mojave Desert in the Miocene epoch, about 20 million years ago. The faulting was of a peculiar kind in which the crust stretches dramatically. Death Valley is stretching like this today. Basins such as Rainbow Basin (Vignette 13), produced by this extension, filled up with lakes and sediments.

Since the big Miocene blast, much of southern California has been geologically quiet except for earthquakes related to the San Andreas fault system (Vignettes 11, 13) and the occasional volcano (Vignette 14). Eastern California, around Death Valley and the Owens Valley, is still stretching, and exhibits great relief, frequent earthquakes, and active volcanism (Vignettes 19, 20).

Geological Vignettes

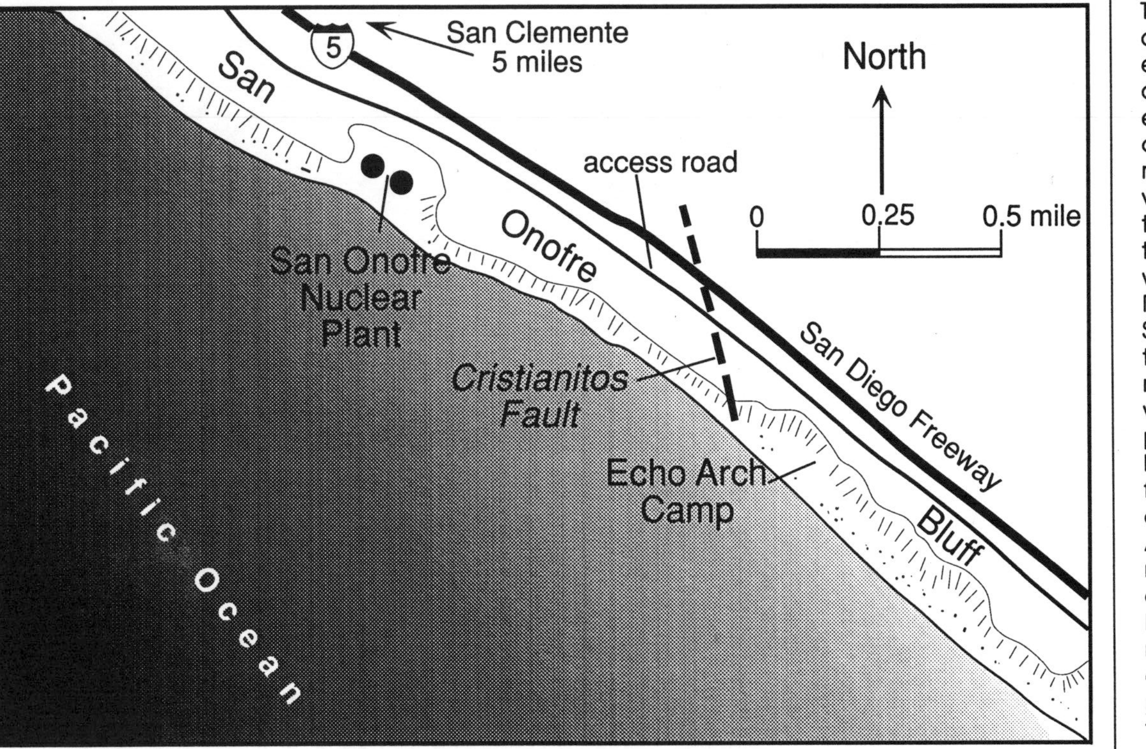

Features at San Onofre nuclear plant.

The reactor domes two miles south of the Basilone-San Onofre Road exit are a familiar sight to the hordes of people who travel the San Diego Freeway (Interstate 5). Details of this area appear on auto club maps and the U. S. Geological Survey, 7.5-minute, San Onofre Bluff topographic quadrangle. To get there, turn off the San Diego Freeway at the Basilone-San Onofre Road exit and follow signs along San Onofre Road, first west and then south, to the state park. This route follows the old Coast Highway past the nuclear plant. A state park entrance kiosk lies 0.8 mile beyond the plant and three miles from Basilone exit. Ask the kiosk attendant for the map of Echo Arch Campground. Go about 0.2 mile from the kiosk into the parking area beyond the first restroom. Trail Number One provides the easiest route to our destination. The best entry to it is at the gated service road nearly two-thirds the distance through the parking area toward restroom two. Follow this road north through Echo Arch Campground all the way to the beach. The distance, about half a mile, requires about fifteen minutes. The climb out ascends 150 feet and may take twenty minutes.

Vignette 1

San Onofre Nuclear Reactors and the Cristianitos Fault

San Diego County

People worry about nuclear power plants. Among other things, they worry about the physical condition of the plant site, because that is critical to its safety. The nuclear power plant at San Onofre Bluff, on the coast south of San Clemente, stands close to the Cristianitos fault, and that raises serious questions. If that fault were to move, would it cause immense earthquake damage, perhaps endangering the power plant? The threat earthquakes normally pose to nuclear power plants is relatively low because the structures are so sturdily built. A distinguished and highly regarded earthquake engineer considers the reactor domes at the San Onofre plant the most earthquake-resistant structures in California and says they are the safest place anyone could be during a major earthquake.

Nevertheless, people do worry that the Cristianitos fault might move—possibly enough to damage the San Onofre nuclear power plant, perhaps spreading radioactive contamination over a wide area. These concerns are reasonable if the fault is still moving, or active. In general, faults that have a history of fairly recent movement are considered active; those that have not moved for tens or hundreds of thousands of years are considered inactive. Is the Cristianitos fault active? We will examine the geologic evidence. The best place to see the relationships, without trespassing, is from the beach at the foot of the bluff at San Onofre State Beach Park, down the Echo Arch trail.

The Echo Arch Campground occupies a curious, knobby bench that interrupts the otherwise nearly vertical San Onofre Bluff. This bench lies

Air photo looking east to San Onofre nuclear generating plant, San Diego freeway in the background. —courtesy Philip J. West

within a semicircular amphitheater that indents the bluff face. A nearly vertical, concave cliff 80 feet high, Echo Arch, bounds its inland side. The campground is on an old landslide that caved in after the waves undercut the sea cliff. Echo Arch is the breakaway scarp, the exposed part of the slip surface on which the landslide moved. Be grateful for the slide; without it, we would have to descend a 140-foot cliff to reach the beach.

Continue through the camp to the beach and walk another 100 feet northwest on the trail along its edge to a clear view of the bluff and the San Onofre power plant. The geologic story of the Cristianitos fault is revealed in the sea cliff.

The lower 30 to 40 feet of the cliff exposes white rock, massive sandstone that rises to a smooth, even, nearly horizontal surface. Lying on that smooth surface, on top of the white sandstone, is a bouldery layer with nearly 100 feet of dark brown gravel and dirt above it.

The smooth upper surface of the white sandstone is an old wave-cut platform, now buried under the bouldery layer and brown gravel. The waves are now cutting a similar platform as they eat away at the present sea cliff. Currents, swells, and breakers attack and erode the shoreline, undercutting the cliff until it caves in. The sea then clears the rubble away, and again undercuts the cliff until it again caves. This continuing

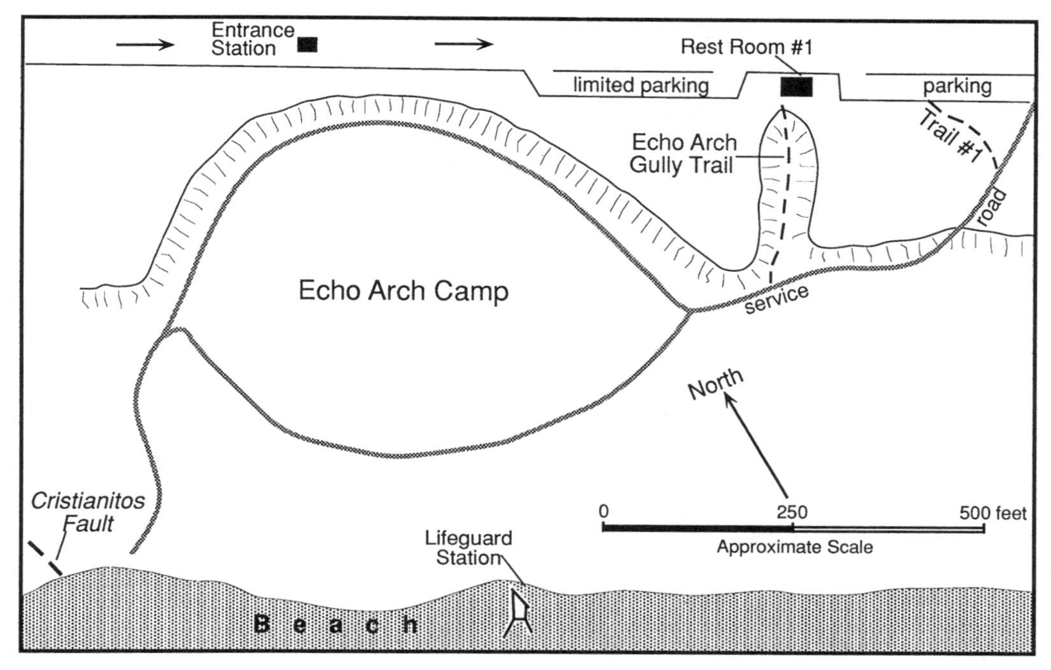

Access routes to Echo Arch campground and Cristianitos fault.

cycle creates a low, gently seaward-sloping plane—the wave-cut platform—that grows wider in the wake of the retreating sea cliff.

Waves must cut those platforms below the level of high tide, but this one is now 40 feet above high tide. Either the land rose or sea level dropped, or both. Relationships elsewhere suggest that sea level was about 20 feet higher when the emerged platform was cut. But between here and Dana Point, 14 miles up the coast, this platform rises by nearly 90 feet. That tilt means the land has risen. So about half the emergent height of the platform is probably due to a drop in sea level, about half to a rise of the land.

Look northwest along the trace of the emerged wave-cut platform buried under boulders and brown gravel. You see that the nuclear plant stands on a foundation made by removing 60 feet of the brown deposits, and excavating about 20 feet into the sandstone beneath the old wave-cut platform. This smooth bedrock surface provides an excellent foundation, stable and sound. Construction engineers should appreciate all the natural processes that combined to provide the emerged wave-cut platform.

So, what is the dark-brown deposit, 60 to 90 feet thick, that buries the wave-cut platform? Imagine what would happen here, if tomorrow sea

Sea cliff 0.8 mile southeast of nuclear plant shows, from bottom up, foreground beach, rilled cliff of white San Mateo sandstone that is smoothly truncated by horizontal wave-cut platform, overlain by thin marine boulder layer, topped by thick, layered, terrestrial deposits. The cliff is 120 feet high.

level were to start falling, or the land rising. In either case, the shoreline would move seaward, exposing the modern wave-cut platform. The beach would move progressively seaward across it, leaving a thin layer of sand, beach stones, sea shells, cans, bottles, sandals, and what-not. Streams, slides, and mudflows would carry rocky, dirty, weathered debris from the inland hills behind the abandoned sea cliff onto the emerging platform. This material would accumulate initially on the platform's landward edge, building an apron sloping seaward. As this apron thickened and steepened, it would also grow seaward because streams could carry debris across its sloping surface to the outer edge. All of that material would be the equivalent of the boulders and brown gravel that now cover the old wave-cut platform.

It might take a good part of 100,000 years to accumulate the ninety feet of debris exposed in the present bluff. Once sea level stabilized, waves would again erode the land, carving a new wave-cut platform and creating the present sea cliff, revealing the record we see exposed in it. This sequence of events has happened repeatedly along the southern California coast, creating a succession of emerged wave-cut platforms and abandoned sea cliffs. The emergent wave-cut platforms and the nearly flat expanses of old sea floor beyond them are called marine terraces. Four easily recognized marine terraces border the coast between San Onofre Bluff and Oceanside, like a flight of giant steps rising from the beach inland into the hills. As many as twelve terraces notch the Palos Verdes Hills to the north.

So, the San Onofre nuclear power plant stands firmly seated within an emerged wave-cut platform, above a twenty-foot-high sea cliff. In the natural course of events, during centuries or perhaps millennia, waves

San Onofre nuclear plant, viewed from southeast, sits on a foundation excavated into massive San Mateo sandstone below the level of the marine wave-cut platform. The ribbed sea wall is one of two concrete embankments protecting the plant from wave erosion.

would undercut that cliff, causing it to slide. But two solidly built concrete walls with an embankment of large blocks of rock at their base protect it from the waves. They will not bring the plant down for a very long time.

A closer look at the sea cliff will reveal details of the Cristianitos fault. The white sandstone lying northwest of the fault is part of the San Mateo formation, which was laid down during Pliocene time, perhaps 5 to 4 million years ago. The brown rock southeast of the fault is part of the Monterey formation; it was deposited during Miocene time, about 20 to 15 million years ago. Ages of the formations show that the northwest side dropped with respect to the southeast side, bringing the younger San Mateo sandstone against the older Monterey formation. The slightly curved fault plane is inclined about 60 degrees to the northwest, but it is much too weathered to preserve direct evidence of movement, such as scratches or smoothed and polished surfaces.

The Cristianitos fault is a major structure; it extends inland on a bearing slightly west of north a projected distance of about 26 miles, crossing Ortega Highway (California 74) and extending up Cañada Chiquita. Fault movement produces earthquakes, and here is a beautiful fault, just 0.9 mile from a nuclear power plant. Should we worry?

Southern California has many faults—of the geologic variety. Some are active, but more are dead. Is Cristianitos fault active or dead? Take a close look at the fault. You will seldom have better opportunity to put your finger right on one.

Smaller, sympathetic faults of similar orientation cut the white beds of San Mateo sandstone for 120 feet to the northwest. Many still smaller faults with three to six inches displacement also break the sandstone layers, which tilt toward the fault at about fifteen degrees. Layers of the Monterey formation, largely dark-brown shale that weathers grayish with some white layers of fine sandstone, are severely broken and crumpled close to the fault. What looks like bedding in this material is shear banding parallel to the fault.

The old wave-cut platform slices right across the Cristianitos fault, but shows absolutely no evidence of displacement. The boulder layer just above the platform as well as the 90 feet of overlying dark-brown alluvial deposits are likewise not displaced. Obviously, the fault has not moved since the waves eroded the platform, and the overlying deposits buried it. That is good news, but how old is the wave-cut platform?

Careful studies suggest that the wave-cut platform is probably about 120,000 to 125,000 years old. Good engineering practice, depending upon the agency involved, requires that a fault that has moved in the last 11,000 to 35,000 years be considered active. The Cristianitos fault is clearly beyond those limits.

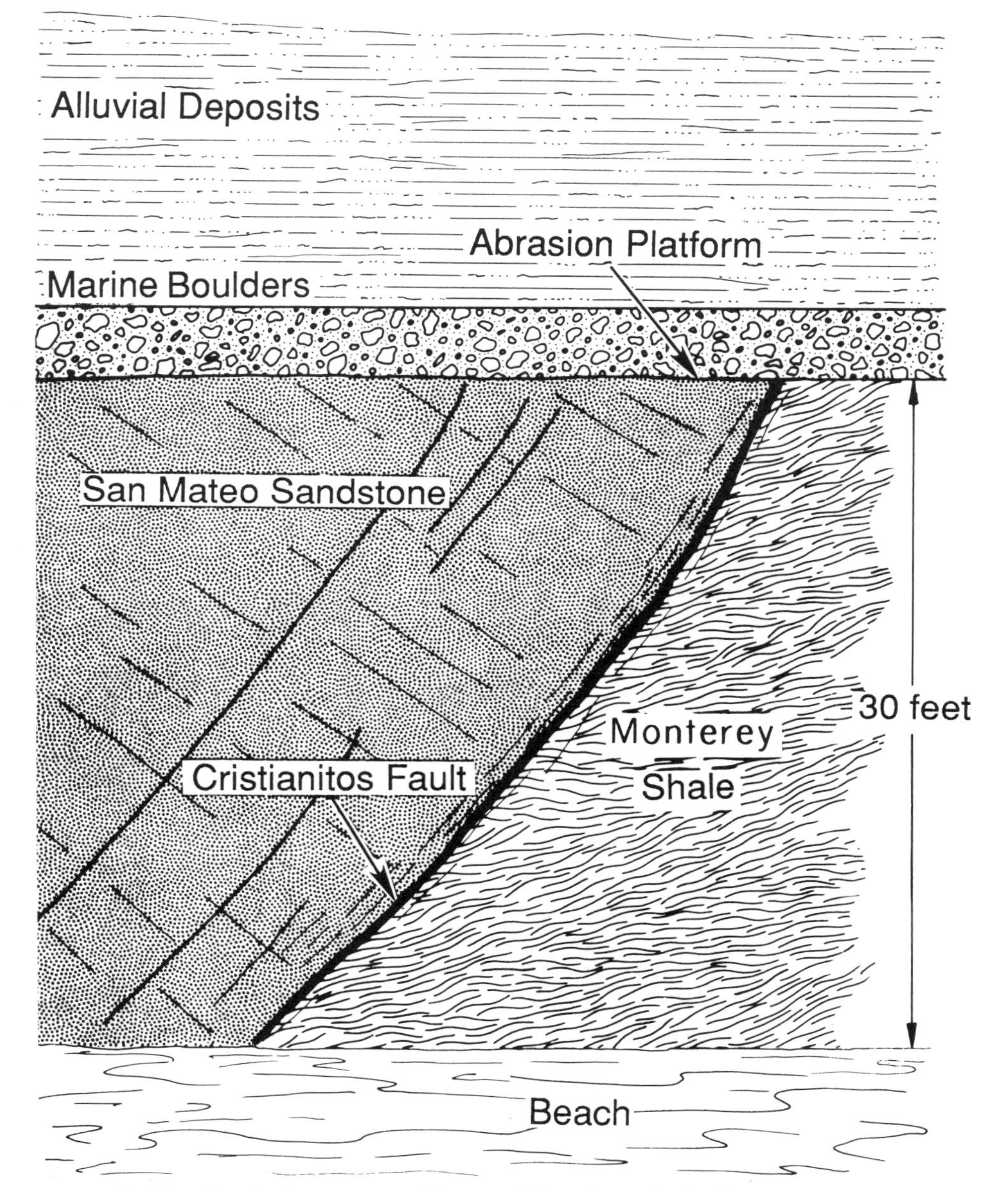

Cristianitos fault as exposed in sea cliff at the end of Echo Arch camp trail; compare with the photo on page 16.

Three other prominent faults parallel to the Cristianitos fault cut the coast between San Onofre Bluff and Dana Point to the north. None shows evidence of movement in the last 125,000 years. Evidently, the stress system that moved them is no longer functional.

How reliable is the 125,000-year age, and on what is it based? As the shoreline receded westward across the now-emerged platform, it left behind a layer of nearshore deposits including fragments of fossil sea shells. Analyzing chemical changes among amino acids within the shells provides one way to approach the age problem, but a better date for these

The Cristianitos fault is visible in this sea cliff, 0.9 mile southeast of nuclear plant. San Mateo sandstone lies to the left of the inclined fault plane, Monterey formation shale to right. A wave-cut platform, overlain by a layer of marine boulders, truncates the fault. The fault does not displace platform, marine boulders, or terrestrial alluvium at top.

shells can be determined by analyzing for the daughter products resulting from the radioactive decay of the tiny amount of uranium within them. Those results suggest an age of about 125,000 years. This procedure works especially well with fossil corals. Fossil coral is rare on emerged southern California terraces, but the Nestor terrace at San Diego contains coral fragments that give consistent dates at around 125,000 years. Is the wave-cut platform at San Onofre Bluff the same as the Nestor terrace at San Diego? Careful geological work confirms that the platform at the nuclear plant correlates with the Nestor terrace of San Diego. Furthermore, fossil shells from the San Onofre platform independently give isotopic ages centering around 120,000 years.

So, you need not worry that an earthquake originating on the Cristianitos fault will damage the San Onofre nuclear generating plant. That fault has not moved for at least 125,000 years, perhaps not since long before then. It is dead.

Faults offshore may be more of a threat. One in the San Diego trough about thirty miles west to southwest of Oceanside caused a magnitude 5.3 earthquake in 1985, but that was too far away to pose a serious threat. Geologic maps and some geophysical data suggest that a southern extension of the Newport-Inglewood fault, which caused the magnitude 6.3 Long Beach earthquake in 1933, may lie a few miles offshore from San Onofre Bluff. If so, it might pose some threat.

What is the lesson of San Onofre Bluff? Good exposure of simple geological relationships can provide information of real value to society.

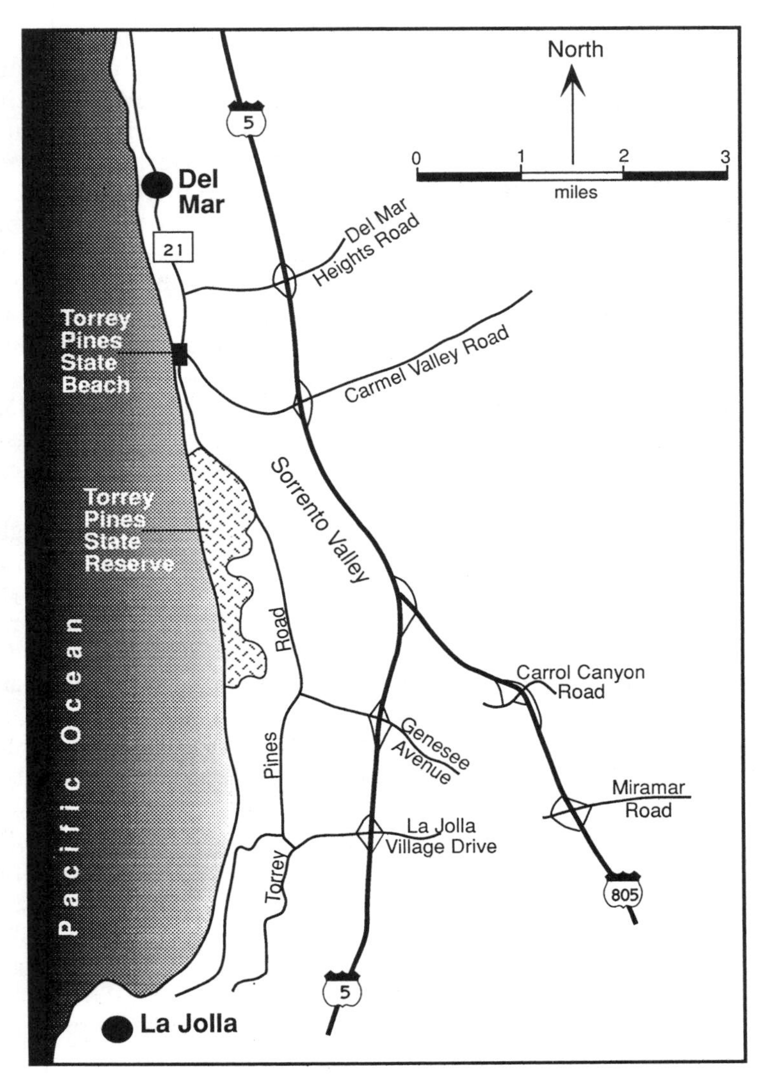

Highways and freeways adjoining Torrey Pines State Reserve and Beach.

Maps of greater San Diego and the U. S. Geological Survey, 7.5 minute, Del Mar topographic quadrangle cover the area. To reach Torrey Pines, drive six miles north from downtown La Jolla on North Torrey Pines Road or 2.5 miles south from the center of Del Mar on Camino Del Mar, both parts of county road S-21. Freeway travelers coming south can turn off Interstate 5 at either Del Mar Heights Road or Carmel Valley Road to reach S-21 north of Torrey Pines. Northbound travelers on Interstate 5 do best to take Carmel Valley Road exit and go west to S-21, then south.

VIGNETTE 2

A MIGRATING SHORELINE

The Story in the Bluffs at Torrey Pines State Reserve

SAN DIEGO COUNTY

Torrey Pines State Reserve and Beach occupy an attractive section of the San Diego coastline. The pines appear to be an ecological relic of the cooler and more moist conditions of the last ice age, about 10,000 years ago, when they were widely distributed along the coast. Under the warmer and drier conditions of our time, only small clumps of Torrey pines survive in favored spots, such as Torrey Pines Mesa. But we will leave those topics to the botanists and ecologists. Our concern is more with the story revealed in the layers of sedimentary rocks exposed in the face of the bluffs.

Torrey Pines bluffs rise abruptly 300 feet above the ocean to the west and are bounded on the northeast by Soledad Valley and on the east by the Sorrento Valley. You may wonder about the wide, flat, swampy floor of Soledad Valley. The enormous glaciers of the great ice ages withdrew enough water to lower sea level several hundred feet. A stream eroded the floor of Soledad Valley to meet the lower sea level. As the melting glaciers returned their water to the ocean, sea level rose, and flooded the lower part of the valley, creating an estuary like those farther north, toward Oceanside. Streams flowing into the Soledad estuary filled it with sediment, transforming it into a swampy valley, its floor hardly above sea level.

Most southern California beaches change dramatically with tide, season, and storms. The long summer swells move to this coast from distant storms. These long waves "feel" the bottom in deeper water than do short waves, so they break farther off shore and generate a swash that carries sand onto the beach. In winter, local coastal storms create short waves that break directly on the beach, and their backwash carries sand away. Thus, sandy beaches usually erode in winter, exposing pebbles and rocks that you rarely see in summer when the waves return the sand to the beach.

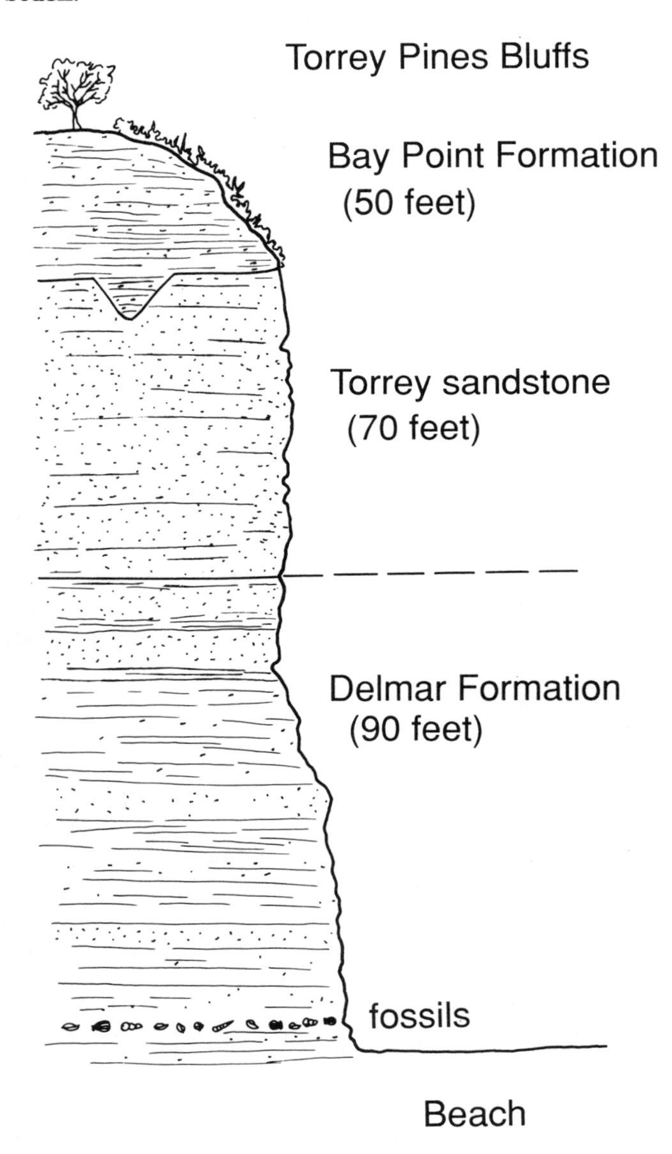

Formations exposed in Torrey Pines bluffs.

Stand near the north end of the bluffs, but near the water's edge, so you can see their full height. Notice that they expose essentially horizontal layers of sedimentary rocks. Most are 45 to 50 million years old. In California, rocks that old are commonly folded, so their layers tilt. The horizontal layers here testify to the rigidity and stability of this small part of the earth's crust, the San Diego block.

Now look at the shape of the bluffs. A nearly vertical cliff rises from their base to within about 50 feet of the top. The uppermost part of the bluff rises as a steep slope beneath a cover of grass and brush to a rounded summit. The Bay Point formation, a relatively young sequence of sedimentary layers, forms this upper slope.

You can see good exposures of the Bay Point beds farther south in the uppermost part of the bluffs, where they make bare slopes much

Torrey Pines bluffs, about 220 feet tall. Dark, bushy slope at top is underlain by Bay Point formation, resting on cavernously weathered cliff of massive, light-colored Torrey sandstone. The partly vegetated, steep slope below this is Delmar formation, well layered in the lower half, massive sandstone in the upper half.

Disconformity of dark Bay Point formation (Pleistocene) filling broad swale cut through Torrey sandstone into top of well-layered Delmar formation (Eocene).

dissected by rills and gullies, miniature badlands. Most of the layers are dark brown clay, silt, or sand so weakly consolidated that you can dig them with a shovel. Locally, the Bay Point formation contains pods and thin layers full of fossilized sea shells, mostly various types of clams. The formation is probably no more than 120,000 years old.

Below the brushy slope on the Bay Point formation is a section of massive and homogeneous rock exposed in a cliff about 60 feet high. It is only faintly layered, and its surface is weathered into an elaborate complex of hollows, niches, chutes, and small caverns. This is the Torrey sandstone, which at about 45 million years is much older than the Bay Point formation. It looks at first glance like a brownish rock, but a bit of digging and scraping would show that it is actually white. Brown debris coming down from the Bay Point formation above stains and coats its surface.

The contact between the Bay Point formation and the Torrey sandstone represents a major gap in the rock record, a long passage of time in which no sedimentary layers accumulated. Here that contact is smooth and level; farther south you will see that deep gullies have been eroded into and even through the Torrey sandstone before the Bay Point formation was deposited.

Beneath the Torrey sandstone is the Delmar formation, which is between 45 and 50 million years old. It makes an irregular but steep cliff, 80 to 90 feet high, containing layers of mudstone, shale, and white or gray sandstone, which lie on a distinctively greenish shale.

Now walk to the base of the bluff for a closer look at the lower part of the Delmar formation and at blocks of rock that have fallen to the beach from layers higher in the sequence. Work south along the cliff base, watching for massive beds of gray rock that are harder than the rest. These beds become prominent ledges within a few hundred feet, and blocks fallen from them litter the cliff base. Look carefully; these rocks contain

Features and locations, Torrey Pines bluffs.

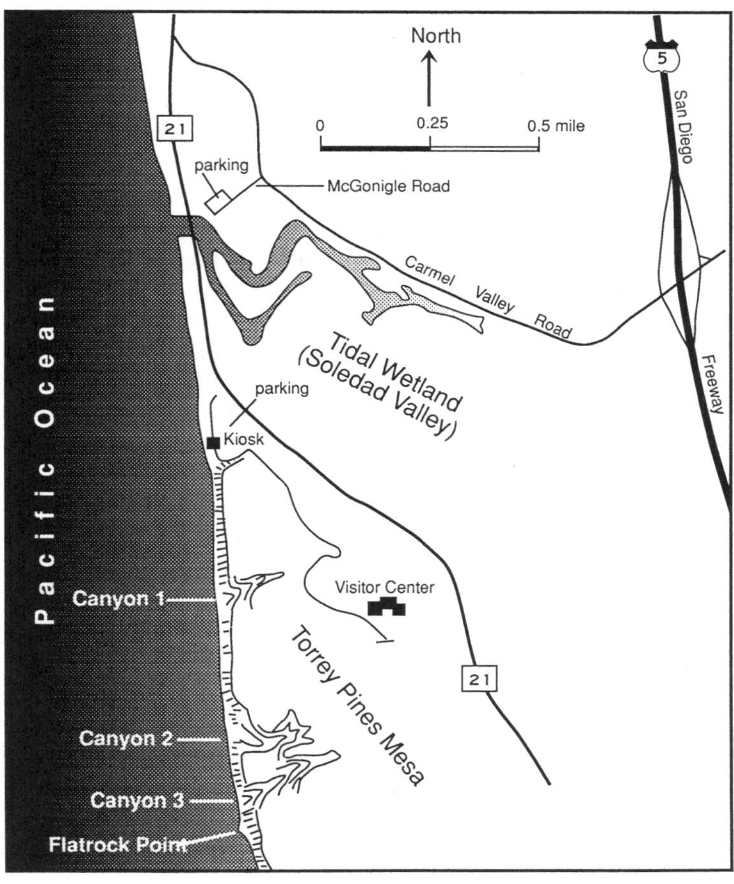

numerous fragments of fossil shells. In some sections, fragments are so abundant that geologists call the rock "fossil hash." Many of the shell fragments are gray, thick, and finely laminated. They are fragments of oysters. Thinner, more fragile pieces of clam shells also abound. Layers of mudstone contain coiled snail shells, some long and elegant spirals. The farther south you walk, the more fossils you see in the Delmar formation, and the better their preservation.

As you stroll along, keep track of the major canyon mouths you pass. If you walk as far south as Flat Rock, you will pass three large canyons: canyon 1 is about 0.4 mile south of the entrance kiosk; canyon 2 has a sign warning you to keep out; canyon 3, just north of Flat Rock, has wooden steps just south of its mouth that lead to a trail and some interesting rocks.

The horizontal layers of Torrey sandstone and Delmar formation on the bluffs are the seaward edges of sheets of sedimentary rock that extend several miles inland. Normally, such sheets are laid down in successive layers, one on top the other, over a broad area. That makes the bottom layer the oldest, the one on top the youngest. Not so here.

About 50 million years ago, this coast was a low plain, and shallow water stretched far offshore. Waves break where they get into shallow water, so the line of breakers was far offshore—it would have been a good surfing spot. The breakers scoured sand from the sea floor and piled it into a broad sandbar. The bar may have risen above sea level to become an offshore barrier island, enclosing a shallow lagoon between itself and the coast. Finer mud and silt accumulated in the calm lagoon. That ancient coast probably looked much like the modern coast of Texas, with its barrier islands and lagoons. Inlets through the barrier island let water pass in and out of the lagoon as the tide rose and fell.

Occasional storm waves, especially at high tide, swept over the barrier island, spreading a layer of sand across the mud in the lagoon. Flooding streams from the mainland also carried sand into the lagoon. Plants and animals generally abound in lagoons, so sediments deposited in them tend to be richly fossiliferous.

The Torrey sandstone looks like a barrier island or offshore sand bar deposit, the Delmar formation like lagoonal muds. But how did the sand bank come to lie on top of the lagoonal mud? Enter the concept of a migrating shoreline.

As long as sea level remains stable, a barrier island and its lagoon remain reasonably fixed in position. But if sea level rises, or the land sinks, the shoreline moves inland. And dropping sea level, or rising land, moves the shoreline seaward. As a shoreline moves inland, its offshore barrier island and lagoon move inland as the deepening water allows the waves to drive the barrier island into the lagoon. Meanwhile, the inner edge of the lagoon also moves inland as the water level rises.

As waves attack the seaward side of the barrier island, they wash sand into the lagoon, at least along its seaward edge. This probably accounts for the massive sand beds in the upper half of the Delmar formation. This movement of bar and lagoon is so slow that a considerable thickness of lagoonal deposits can accumulate before the bar sands arrive on the scene. In due time, the barrier island moves landward across the old lagoon, burying the lagoonal muds under sand.

Most of us have seen the large contraptions that spread concrete or asphalt for a modern highway. The paving machine lays down a ribbon of material of uniform thickness from one end of the project to the other. The farther part is older, the near part younger. Now imagine two paving machines many miles wide, one following the other, each laying down a sheet of sediments across a smooth plain. That, basically, is what happened in the Torrey Pines area 45 to 50 million years ago.

First, rising sea level, or sinking land, moved the shoreline several miles inland, and the paving machine of the transgressing barrier island laid a sheet of sand over the lagoonal mud. Then sea level dropped, or perhaps the land rose. As the shoreline moved seaward, the paving machine of the lagoon spread mud over the layer of sand the barrier island had left as it moved seaward. Few places in North America demonstrate the inland movement of this paving machine mechanism better than the bluffs at Torrey Pines.

Near the steps of the trail up Canyon 3 is a nicely exposed angular unconformity within the Delmar formation. An angular unconformity is a surface along which sedimentary layers lie at an angle to each other, instead of being parallel. The angular discordance here is about ten degrees, between a horizontal massive sandstone bed and the tilted, thinner sandstone and shale layers beneath it. The sandstone and shale layers, one of which is full of fossils, tilt because they were deposited on the side of a shallow scour channel. Erosion truncated the tilted beds, leaving behind a flat surface, before the sandstone was laid down over them. Most angular unconformities separate formations, but this one is a minor local phenomenon within the Delmar formation.

A climb of 100 feet up the trail into Canyon 3 provides a good introduction to the Bay Point formation. The initially narrow canyon shortly opens into a spacious amphitheater set within badland slopes intricately eroded into the Bay Point formation. The Bay Point beds filled an ancient valley cut through the Torrey sandstone and into the Delmar formation. The stream that eroded Canyon 3 created the wide amphitheater as it cleared most of the soft rocks of the Bay Point formation out of that old valley.

Some thin layers of soft, white sand in the Bay Point will catch your eye. But watch particularly for pods, lenses and thin beds with numerous

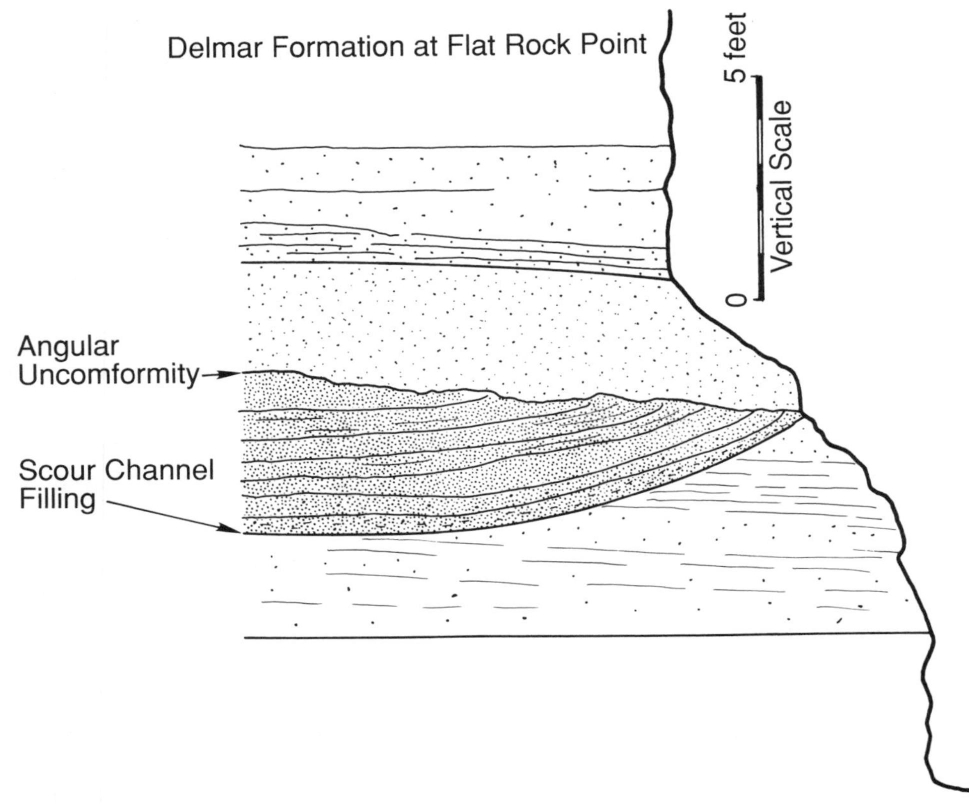

Sediment-filled scour channel and angular unconformity in Delmar beds near Flatrock Point.

white, fossil sea shells that look as fresh as the shells on the present beach. Fragments of clam shells are most common, but bits of snails, sand dollars, worm tubes, and barnacles also abound.

Events of Eocene time, about 35 to 55 million years ago, made the Delmar formation exposed in the lower part of this canyon messy and confusing. The walls of an ancient submarine canyon eroded into the Eocene sea floor collapsed, jumbling the soft sediments into a new deposit—a massive, brownish sandstone with large leopard spots of pale rock within it. Look for it in the upper part of canyon 3.

As you walk back to the entrance kiosk, keep well out on the beach to enjoy broad views of the bluffs. You will see how the brown and intricately eroded slopes of the Bay Point formation differ from the cliffs beneath, which expose the much harder Torrey and Delmar formations.

Nicely smoothed pebbles litter the beach, especially in winter. They are hard rocks—most are volcanic—that resist both the mechanical wear of the waves and the chemical corrosion of sea water. They have been through the mill several times, and have survived. Many are nice little disks, quite different from the more spherical shape typical of stream rounded pebbles. Waves sliding and rolling them on the sandy beach wore them to that shape, which is common in beach pebbles.

These smooth pebbles may too become part of the geologic record. As changing sea levels alter this coastline, deposits preserved from this sandy beach and the distant estuaries may provide glimpses of the landscape we see today.

Access map for Aliso Beach, South Laguna.

To get there, take California 1 south from Corona Del Mar and Laguna Beach, or north from Capistrano Beach. An auto club map of Orange County and the U.S. Geological Survey, 7.5-minute, Laguna Beach topographic quadrangle are useful.

Vignette 3

A Vanished Landmass

San Onofre Breccia at Aliso Beach

Orange County

The San Onofre breccia (pronounced "bretch´-yuh") is one of the most unusual formations in southern California. It is nicely exposed in the sea cliff at Aliso Beach County Park, in South Laguna. The late A. O. Woodford, a Pomona College professor who lived to the ripe age of one hundred years, first described the deposit in 1925, in a classic paper.

Among geologists, a formation is a body of rock that they can recognize from one outcrop to another and follow through a fairly large area. Breccia is a mixture of sharply angular fragments of rock embedded in a matrix of finer particles. The fragments may range from a small fraction of an inch to blocks the size of a bus. They may consist of a single type of rock or of many types, as in the San Onofre breccia formation.

The San Onofre breccia consists of rock debris shed from a steep landmass that stood somewhere nearby. That highland is gone now, its angular fragments preserved in the breccia. Fossils of sea creatures within the breccia show that some of that debris accumulated in the ocean. Most of it contains no such fossils, and appears to have accumulated on land.

To judge from the fragments preserved in the San Onofre breccia, the vanished landmass was a complex assemblage of metamorphic rocks that had recrystallized at only modestly high temperature and under intense pressure. Streams, rock slides, rock falls, and watery debris flows carried rock fragments from the landmass and deposited them in the San Onofre breccia. That happened between 15 and 20 million years ago, when the Laguna coast was quite different from what we see today.

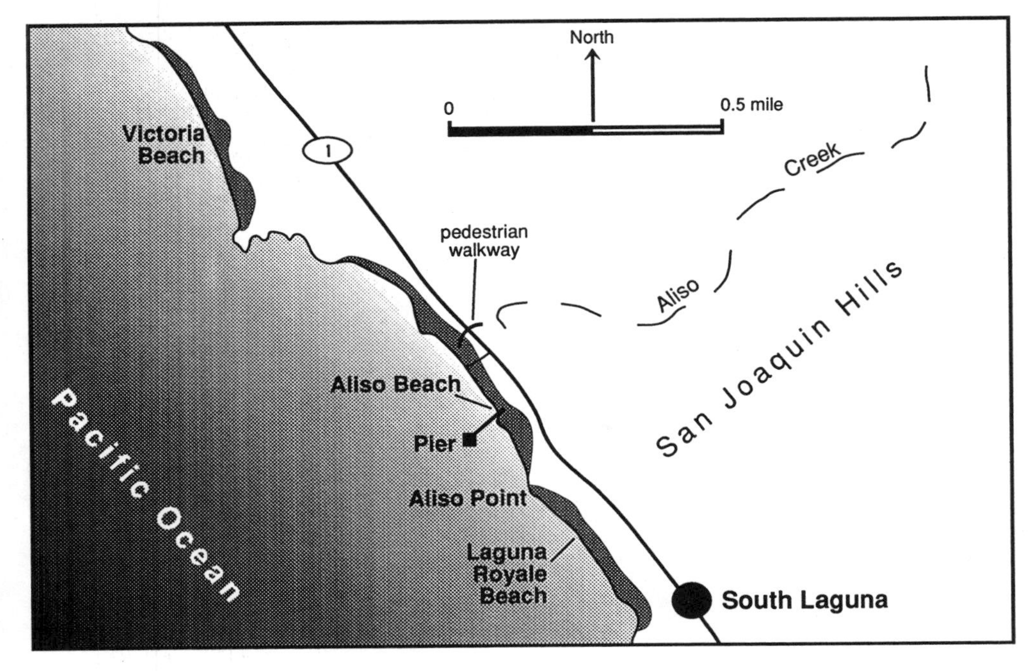

Features at Aliso Beach.

Aliso Beach is one of the nicest of the many attractive cove beaches on the Laguna coast. The sand is coarse and pale, with a scattering of dark grains. Aliso Creek nourishes this beach with a goodly supply of sand carried from the foothills of the Santa Ana Mountains.

From the edge of the parking area, walk southeast down the beach for about 250 feet to the first projecting cliff. It exposes a sandstone within the San Onofre breccia. Notice all the hollows and niches in the higher part of the cliff. You commonly see such cavernous weathering in sandstone exposed in a sea cliff, but rarely see it in the same formation where exposed inland. This distribution suggests that salt spray is somehow responsible for the peculiar weathering patterns, but geologists differ in their ideas about how it works. Some suspect that salt crystallizing within the rock breaks it apart; others attribute the weathering to more complex chemical processes.

Walk another 100 feet to the next projecting cliff with several large blocks of fallen rock along its base. The deposits exposed in this cliff are faintly layered; see if you can recognize the beds. They tilt as much as 60 degrees from horizontal, which is much steeper than any possible angle of initial deposition. These rocks have been deformed.

Case-hardened sandy phase of San Onofre breccia with cavernous weathering; footprints in sand give scale.

The different layers of breccia here contain fragments of different sizes. Some are as small as pebbles, less than two inches across, and some as large as boulders, greater than ten inches across. All are embedded in a sandy groundmass of finer rock particles.

Look carefully at individual rock fragments. Most are angular with sharp edges, although some edges are blunted by the wear during transport. Fragments vary in color, texture, and composition, which means that they came from an area that contained a variety of rock types. The dominant rock is a nearly black schist full of nodules that make little bumps on its surface. Schist is a metamorphic rock that has a distinctly laminated structure, as though it were a stack of thin sheets; geologists say it is "foliated" (from the Latin "folium," meaning leaf).

Fragments of a greenish schist abound. It is greenstone, metamorphosed volcanic rock. Chunks of white quartz are probably pieces broken

San Onofre breccia on inland side of Aliso Point, with clastic dike right of center. The largest rock in the breccia, cut by a white calcite veinlet, is 30 inches long.

from quartz veins. Tan to light gray fragments of igneous rocks are only weakly foliated. The foliation of the parent rocks explains why most of the breccia fragments have a platy shape.

Watch here and farther along for platy fragments of a fine-grained and shiny rock, in which good light reveals a distinctly lavender or purplish color. This is glaucophane schist, easily the most distinctive rock among the San Onofre breccia fragments. Glaucophane is an uncommon mineral with a lavender to bluish color and a high sodium content. It forms under very high pressure, but at a relatively low temperature—for a metamorphic mineral. Very few places deep enough within the earth to provide the very high pressure that glaucophane needs to form are also cool enough to permit it to form. Glaucophane almost certainly forms as rocks recrystallize in an oceanic trench, where rapidly sinking oceanic crust carries rocks to great depth before they can soak up much heat. The glaucophane in southern California is in the Franciscan rock assemblage.

The Franciscan is an assortment of rocks that was drawn down into a trench that existed until about 25 million years ago along what is now the southern California coast.

Continue your stroll down the shore to Aliso Point, a ragged cliff with a prominent knob. The rocks in the knob, and in the outcrops inland from it, contain two types of veinlets, thin tabular deposits that fill cracks in the breccia and cut through some of its rock fragments.

One group of veinlets is white and homogeneous calcite, calcium carbonate. You can scratch these veins with a knife and thereby demonstrate they are not quartz. They were deposited by ground water circulating through the breccia after it was firm enough to fracture. On the face of the first outcrop inland from the knob, the whole side of one veinlet has been laid bare, so it looks like the exposure was coated with calcite.

Veinlets of the other group are tan or gray, thicker than the calcite veinlets, and more irregular. They are composed of sandy material like the matrix of the breccia. Some are thick enough to include small rock fragments. One on the south face of the knob of Aliso Point looks like a vein of concrete grout injected into the breccia. These veinlets are more

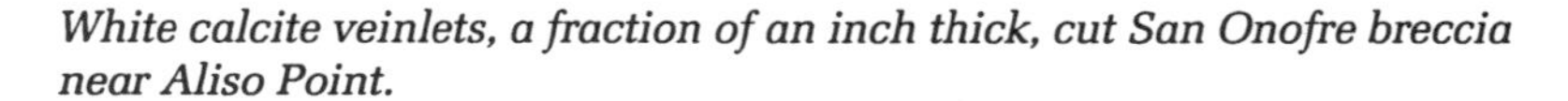
White calcite veinlets, a fraction of an inch thick, cut San Onofre breccia near Aliso Point.

Coarse bouldery phase of San Onofre breccia composing Aliso Point. The 6-to-8-inch-wide vertical feature at the left margin is a clastic dike, a sedimentary dike squeezed in place from overlying or underlying beds.

properly called clastic dikes. A dike is a tabular body, discordantly intruded into a host rock, usually along a fracture. Clastic simply means composed of broken rock or mineral fragments. These clastic dikes were probably injected, perhaps with the aid of an earthquake, when the rock was deeply buried and under high pressure.

Venture around Aliso Point to the next cove beach to see the huge boulders in the breccia exposed in the sea cliff. Some are as large as a small automobile. Look at the cliff face just south of Aliso Point to see a planar band cutting through the breccia, stained brown and inclined about 60 degrees toward the ocean. Beds on opposite sides of this band do not match, although they are both part of the same geologic formation. It is a fault, a fracture along which rock masses slipped past each other.

This extremely coarse San Onofre breccia is exposed in a 40-foot-high sea cliff just south of Aliso Point.

The San Onofre breccia is clearly an exceptional deposit in its own right. But its most unusual aspect is the absence of a source area. No parent rock that could have been its source exists anywhere nearby. That, and other considerations, led Professor Woodford to look out to sea for the source rock. Perhaps a high, rugged landmass stood just off the present shore some 15 to 20 million years ago. Some evidence suggests that a fault lies just offshore, and it might have raised the landmass. If its shoreward side were a steep fault scarp, it could easily have shed the rock fragments that became the San Onofre breccia. The right kind of parent rock does exist on Catalina Island, possibly also in other parts of the submerged continental borderland off southern California. Woodford's postulation has endured 65 years of critical scrutiny, and is still held in high regard.

So, where is that landmass today? Have the waves planed it off? Probably not. An offshore, submarine bank or shoal should remain, and none does. The source of those fragments really is gone.

Numerous faults that move horizontally cut through southern California, many with displacements measured in tens to hundreds of miles. Some of those faults also move vertically, so it is possible that one of them raised the source rocks for the San Onofre breccia, then carried them away. Most of the big faults of southern California have right-lateral displacement—if you look across the fault, the opposite side moves to your right. So, if a fault did carry away the source rocks for the San Onofre breccia, they must now lie somewhere to the northwest, up the coast.

How about the Palos Verdes Hills? They do have a fault along their inland side, and they do contain a small exposure of metamorphic rocks of the right kind to make San Onofre breccia. Much more of that rock could exist there, now buried under younger sedimentary deposits. The required distance of displacement, roughly 45 miles, is sobering, but other right-lateral faults in southern California have moved even farther within the last 20 million years.

Aliso Beach is not the only place where San Onofre breccia is exposed, but it is one of the best. The formation appears in sea cliffs and hillsides all along the Laguna coast, as well as in the hills inland from San Onofre State Beach, south of San Clemente. The breccia beds extend inland a few miles, becoming finer grained with increasing distance from the coast, which is consistent with their postulated source offshore.

Vignette 4

A Growing Fold

Ventura Anticline

Ventura County

The kaleidoscopic structures of southern California are still forming, still changing as the earth's crust moves. These crustal movements are largely responsible for the varied landscapes, and for the unwelcome earthquakes. You can see some of that movement strikingly expressed in the Ventura anticline.

Most southern Californians know that a fault is a fracture along which blocks of the earth's crust have moved past each other—up, down, horizontally, or in some combination. Southern California has so many geologic faults that we tend to forget the crust can also deform by simple tilting, warping, or folding. Folds form best in well-layered, weak sedimentary rocks, of which Ventura County has a great thickness.

Bend a telephone book up into an arch. It's easy because the pages slip past one another. Were they stapled together, the book would be harder to bend. It is the same with rocks. Soft, well-layered rocks can fold, whereas strong, massive rocks fracture under stress, creating faults. Faults also form in severely stressed sedimentary rocks.

An arching fold is an anticline, so named because the opposing limbs incline away from each other, thus anti-inclined. Fold the rocks into a trough and you get a syncline, in which the opposing limbs incline toward each other, hence the term syncline, meaning together-inclined.

Oil, gas, and water can collect in anticlines, if those fluids, and porous beds to hold them, exist within the rocks. As these fluids rise to the crest of an anticline, they arrange themselves in order of density: gas on top, then oil, then water—all trapped beneath impervious layers of rock.

Petroleum geologists and engineers generally plan wells to tap into the oil zone, rather than the gas. As much as possible, they preserve the gas below ground because it helps force oil to the surface. Many anticlines are shaped like an inverted canoe with the central line, or axis, along the crest plunging down at both ends. That form securely traps the oil and gas. A few anticlines are cracked so they leak oil and gas; that is the cause of the natural oil pollution in the Santa Barbara Channel and the tar blobs along the shoreline near Goleta and Carpenteria.

About 2.5 miles north of Ventura, along California 33, is a truly beautiful, large anticline that extends east and west for fifteen miles. Dubbed the Ventura anticline, it is the site of a highly productive oil field, one of California's largest with a cumulative production approaching one billion barrels of oil and several trillion cubic feet of gas.

Most folds extend to great depths. Some geologists think the Ventura anticline ends on a horizontal fault at a depth of about 11,000 feet.

California geologists long ago recognized that the Ventura anticline had to be very young, because many of the fossil sea shells in the tilted beds are the same species as those still living along the coast. They estimated the age of the fold as only one or two million years. Geologists

Eastward view along the axis of the Ventura anticline and oil field, with the Ventura River floodplain in foreground. This photo was taken February 15, 1950. —Spence air photo B-1952, Department of Geography, University of California, Los Angeles

in more stable parts of the United States regarded that figure as insanely young. Several kinds of evidence now show that even these early estimates of its age were far too old.

Obviously, a fold must be younger than the youngest beds it bends. Sedimentary rocks deposited in sea water just 200,000 years ago are steeply tilted on the flanks of the Ventura anticline, so the fold must have formed since they were laid down. That is amazing; in most regions folds are tens or hundreds of millions of years old. But terraces along the Ventura River that are only 16,000 years old tilt where they cross the fold. That suggests the fold may still be rising.

The ages of the tilted rocks and the amounts of their tilting show that the Ventura anticline has risen at an average rate of nearly 0.6 inch per year for the last 200,000 years. A little simple arithmetic shows that the crest has been lifted roughly two miles in that time. Rise of the fold crest has now slowed to less than 0.1 inch per year, but that is probably due to the changing shape of the fold, not to a slowing of the folding. As the limbs of a fold steepen, its crest rises more slowly. You can see how that works by laying a sheet of paper flat on a table, then pushing its edges together to make it into an anticline. As the limbs of the fold become nearly vertical, its crest almost stops rising, even though you continue to bend the paper as fast as ever.

Level-line surveys show that the Ventura River valley rose about 0.2 inch per year between 1920 and 1968. Some of that rise might be due to creep on faults, but much of it is the result of folding. Further, the ground above large oil fields typically sinks as oil, gas, and water are extracted. The absence of subsidence at the Ventura oil field indicates the anticline is still going up. No doubt about it, the anticline is still growing. This is one of the few growing anticlines in captivity. Salute it!

The rocks in the Ventura anticline are very soft; you can dig many of them with a shovel. Erosion has already stripped thousands of feet of them from the rising crest of the fold. If it were not so, the anticline would now make a ridge towering many thousands of feet above sea level.

On its way south to the coast, the Ventura River eroded a gap through the rising anticline, which California 33 follows between Ventura and Ojai. The river maintains this transecting course by eroding its bed as fast as the anticlinal ridge rises across its path. Raised remnants of old floodplains of the Ventura River survive as terraces, broad steps above the modern floodplain extending to the valley walls. When the river was flowing on them, these old valley floors sloped as the river flows, gently downstream toward the sea. The rising fold has tilted the terraces on its north limb so they now slope upstream.

Marine terraces on the south limb of the anticline are old wave-cut benches that originally sloped about one degree seaward, the usual slope of such terraces. They are two to three times older than the stream terraces on the north flank of the fold. The marine terraces now slope about ten degrees seaward and stand hundreds of feet above sea level. Part of that rise is due to the folding, the rest to movement along faults.

The thick section of folded sedimentary beds in the Ventura anticline contains several thin layers of volcanic ash. Volcanic ash incorporated in sedimentary rocks is generally altered to clay, and that makes it impossible to date directly by radiometric measurements. However, it is generally possible to fingerprint volcanic ash through chemical analyses and microscopic studies, to match it with volcanic rocks elsewhere, and so find the volcanic field where it erupted. Then it becomes possible to obtain isotopic dates on volcanic rocks associated with the ash, thus fixing its age.

The well-known Lava Creek ash layer lies within steeply tilted beds on the south limb of Ventura anticline. It exactly matches ash erupted during a gigantic volcanic explosion in Yellowstone Park about 600,000 years ago. About 3,000 feet of younger sedimentary rocks above that ash are also tilted. Other less exact dating techniques, and the estimated rate at which sediments are accumulating on the sea floor off California, suggest that the youngest tilted sedimentary layers are about 200,000 years old.

This is all very interesting in an abstract way, but can you actually see some of the relationships? Certainly. Drive north from Ventura on California 33. In passing the first exit, Stanley Avenue, about 1.5 miles north, you see on the east valley wall a large scar excavated in soft shale beds to make drilling mud for the oil field and to provide raw material for an expandable clay plant. The sedimentary layers in the face of the scar tilt down to the south at an angle of about 40 degrees. You are well out on the south limb of the anticline.

A little farther north you can see massive sandstone beds with a gentler southward tilt exposed in the west valley wall. Near the Shell Road exit, sandstone layers tilt even more gently down to the south on both sides of the road. The layers exposed in the cuts on the east valley wall opposite the large sign announcing "Cañada Larga Exit, 1½ miles" are nearly horizontal.

Horizontal makes sense at that place because beds at the crest of an anticline are horizontal, unless the central line of the fold is tilted. The trend of this fold's central line is slightly oblique to the Ventura River and California 33, so the corresponding horizontal beds lie somewhat farther south on the west valley wall. Then, the first beds you see west of the Cañada Larga sign tilt gently down to the north. That means we have crossed the crest of the fold.

Exit at Cañada Larga Road. Stop at the bottom of the off-ramp, then turn west (left) under the overpass to the next stop. Turn south (left), and go south on what becomes Ventura Avenue. You pass the north end of

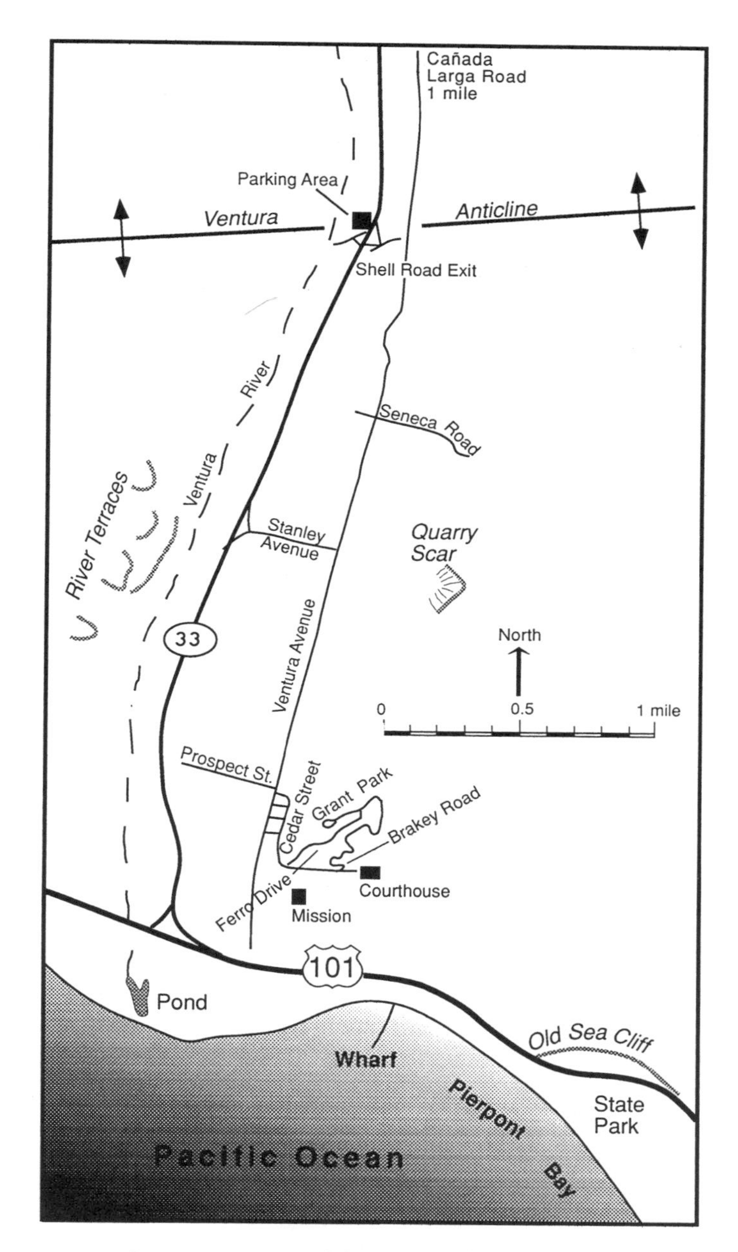

Highways, streets, and features of the Ventura area.

Auto club maps of Ventura County, a city map of Ventura, and the U.S. Geological Survey, 7.5-minute, Ventura topographic quadrangle map are helpful.

Crooked Palm Street on the west (right) just before crossing under the freeway, and you see the south end of Crooked Palm in another 0.9 mile. In another 0.3 mile is a large dirt parking area on the west (right). Park near its south edge. Look in the lowermost part of the east valley wall for a cut bank exposing horizontal sedimentary beds. Another cut on the north wall of the gully just to the north also shows horizontal beds. You are right on the crest of the anticline. If you were on either flank of the fold, you would see tilted layers of rock.

As you continue south, look east beyond Shell Road to see exposures of massive sandstone beds tilting down to the south at an angle of 20 to 30 degrees. You are now well out onto the south limb of the fold. Between Seneca Street and Stanley Avenue, you again see the big cut on the east valley wall, with beds steeply inclined down to the south.

Between 1895 and 1898, Ralph B. Lloyd, a recent graduate from the University of California at Berkeley and a Ventura resident, did essentially what we have just done, but on a horse. He knew what an anticline looked like, and was aware that oil and gas accumulated in such structures. Ralph Lloyd became a dogged advocate of drilling for oil on the Ventura anticline. His was a long, rocky road, but finally on September 25, 1916, Lloyd #2 came in with the first significant oil production. Shallow wells had earlier (1903) produced gas. Lloyd later drilled 57 successful wells before his first dry hole, a record any operator could envy. Ralph Lloyd became the father of the Ventura oil field because he recognized the significance of those tilted beds—and he also became a wealthy man.

Turn left off Ventura Avenue at Ramona Avenue, 0.7 mile south of Stanley, and head for Cedar Street at the base of the hill. Follow Cedar Street to the south (right) and uphill. Near the top, just before Cedar is split by a center divider, turn north (left) onto Ferro Drive, and follow it, by keeping left at all intersections, to Grant Park and Padre Serra's cross. If you miss Ferro Drive, keep going on Cedar, which shortly becomes Poli Street. In a block or so, at the large county courthouse and jail buildings, Brakey Road turns narrowly and steeply to the north (left). Follow it under a large stone arch to an intersection, well uphill, with a sign that directs travelers west (left) to Padre Serra's cross and Grant Park, a superb spot for viewing the geologic features around Ventura.

Walk to the northeast corner of the parking area and descend to the dirt walkway outside the upper hedge and its retaining wall. Follow the walkway counterclockwise around the parking area. Look north into the interior of Ventura County; west and northwest to the Rincon coast. To the west is Santa Barbara Channel and to the southwest, the Channel Islands. To the south and southeast you see the Oxnard Plain and the spectacular beach strand toward Port Hueneme. Point Mugu rises starkly on the southeast skyline, and the ragged, high, western end of the Santa

Monica Mountains is to the east and southeast. Santa Clara Valley lies directly east.

The Ventura anticline, especially its eastern part, is best seen part way around the path. Since drilling rigs no longer remain over wells, as in earlier years, identification of the oil field and anticline is not easy. The crest of the anticline is beneath a wide belt of brush and scattered oak trees that contrasts with the mostly grassy slopes of the limbs on either side. The broad, flat floor of the Ventura River narrows to a gap through massive sandstone beds in the core of the anticline. The oil field has many secondary hillside roads, cut banks at well sites, tanks, and large gas compression plants toward the east end. Not all the hillside scars are artificial. Some are the scarps at the heads of landslides, which have given oil companies fits by shearing off producing wells 100 to 200 feet below the surface.

Look directly up the Ventura River valley and think of that direction as 12 o'clock on a horizontal clock face. Across the Ventura River at 10:30

View northwestward across Ventura River valley, showing remnants of stream-terraces at two levels on west bank.

View south-southeast from Grant Park shows Ventura harbor and curving Pierpont Bay, with groins being the dark structures on the beach.

o'clock you can see remnants of two stream terraces, and at 9 o'clock the smooth skyline profile of the marine terrace that now tilts ten degrees seaward.

Farther around to the south, you can see a pond at the mouth of the Ventura River, Ventura harbor, its pier, and the beach. Offshore is Santa Barbara Channel, and on the far skyline at least two of the channel islands, Anacapa on the east and Santa Cruz on the west. Beyond the steps to Father Serra's cross is a good view of the lovely curving strand of Pierpont Bay. The bay gives way south-southeastward to the long strand extending past the breakwater at Ventura Marina, the power plant with lofty stacks at the mouth of the Santa Clara River, and on toward Port Hueneme.

From 1855 until at least 1933, the beach at Pierpont Bay was building outward into the ocean at an average rate of about seven and a half feet per year. After the Ventura River was diverted for irrigation and flood control, it could no longer deliver sand to the coast. That deprived the Pierpont shoreline of its steady supply of sand, and the waves then eroded the beach some 200 feet inland within 15 years. Twenty-two acres of beach were lost, at a value of $50,000 per acre—that's over a million dollars.

Look along the coast for groins, walls that project from the shore into the surf. The groins are intended to impound sand moving southeast along the beach, and to prevent the loss of more beach. To some degree, they do those things, but the gains they produce become someone else's loss. The sand the groins impound never moves farther down the coast. The structures deprive the beaches farther along the shore of their share of sand.

A healthy stretch of beach is in a state of dynamic equilibrium between the sand that comes in one end and the sand that moves out the other. They balance. That is why a beach can appear unchanged even though sand flows along it as though it were water flowing down a river. Groins, because they disrupt the natural flow of sand, are not good solutions to the beach deterioration that follows river diversion.

Look farther southeast to see Point Mugu and the west end of the Santa Monica Mountains rising abruptly in skyline silhouette above the Oxnard Plain. The relatively smooth skyline extending east from Point Mugu continues a considerable distance before it abruptly rises into the ragged skyline of Old Boney Mountain, a brooding mass that frowns down on the lowland to the west. The lower, smoother skyline is underlain by relatively soft sedimentary beds. Old Boney is held up at a higher and rougher skyline by harder and more resistant volcanic rocks.

The Channel Islands, the westward extension of the Santa Monica Mountains, are part of the continental borderland, the marginal part of the continent that lies shallowly submerged. The true edge of the continent, bordered by the deep ocean, is more than 100 miles offshore. Rocks of the Channel Islands are the same as those of the mainland. Large faults that cut the islands are extensions of mainland faults. Some of the oil-bearing structures in the Santa Barbara Channel—Rincon anticline, for example—are the extensions of mainland structures.

If the air is clear down to water level, you may be able to see the famous sea arch at the east end of Anacapa Island. The arch is actually in a large offshore rock, but from this view it appears to be part of the island.

Look from Point Mugu west through the chain of four channel islands—actually six, because Anacapa is three separate islands. It is easy to imagine that a drop in sea level would convert them into a long peninsula projecting west from the Santa Monica Mountains, converting Santa Barbara Channel into a bay. This hypothetical peninsula has been called Cabrillo Peninsula to honor Juan Rodríguez Cabrillo, the Portuguese-born, Spanish soldier and explorer who sailed the California coast in 1542 to 1543. The official narrative of the expedition, written in 1855 and discovered in the Spanish archives in Seville over three centuries later, say he died about January 3, 1543, on San Miguel Island; there he was probably buried.

The Cabrillo Peninsula probably did exist during the last ice age, when sea level was some 300 feet lower than it is now. It became the present string of islands as the melting glaciers returned their water to the ocean, about 12,000 years ago. The ice age left some surprising souvenirs. The small grove of rare Torrey pines on Santa Rosa Island may be one such memento of the Cabrillo Peninsula. The trees could have migrated much more easily along a peninsula than across a sea strait twenty to thirty miles wide.

The same line of reasoning explains the fossil bones of pygmy elephants on the islands. No trace of similar pygmy elephants is known on the mainland. Some investigators suggest that big elephants swam the channel to reach the islands, then evolved there into a race of pygmies. But what would motivate an elephant to swim twenty miles? It is easier to imagine them walking along the peninsula with something to eat on the way, then finding themselves isolated as a rising sea level converted the peninsula into a chain of islands. Why did they become pygmies? Perhaps because the limited food supply on the islands could not support big animals or because of inbreeding of a small population.

VIGNETTE 5

PERILS OF TAMPERING WITH NATURE
The Harbor at Santa Barbara

SANTA BARBARA COUNTY

Despite its ruggedness, California's coast has few good harbors, mainly because it is rising, which tends to produce straight shorelines without protected indentations. Artificial anchorages, such as the one at Santa Barbara, are harbors only in name, a form of political license.

Greater Santa Barbara has long been favored by people of comfortable means. In early days some owned yachts, and at present many more own small sail boats. Commercial shipping includes a sizeable fishing fleet and excursion boats for travel along the coast and to the channel islands.

Interest in an anchorage here dates back to at least 1850, but little was accomplished until 1926 when Major Max Fleischmann, of yeast fame, offered a sum of $200,000, to be matched by public funds, for construction of a breakwater about 600 feet offshore, parallel to the beach. The site selected was off Point Castillo. This was referred to as the Castle Rock site, because of a ragged offshore rock, later demolished. The breakwater lies a little west of Stearn's wharf, a structure erected in the 1870s and subsequently modified several times. Blocks of rock to build the breakwater were hauled by barge from Santa Cruz Island.

The coast at Santa Barbara trends nearly due east and west, and large volumes of sand are carried eastward along this shore by waves and currents. Some people realized this from the beginning. The longshore sand flow occurs because most of the impinging swells are generated by storms far out at sea to the northwest. Although bent by refraction as they move into shallow water, these swells remain slightly oblique to the beach in an easterly direction. This creates an eastward longshore drift.

Occasional storm waves from the southwest temporarily reverse the normal longshore drift direction.

Construction of the offshore breakwater was complete by 1929, but photographs show that sand soon began to accumulate in the harbor and dog the project. The original expectation had probably been that an offshore breakwater would permit the longshore drift to carry sand on east, leaving the harbor unobstructed. Unfortunately, principles of ocean-wave refraction were not fully understood or appreciated in the 1920s. The east end of the breakwater bent swells westward so they interfered with the normal drift of sand, filling the harbor basin and depositing a broad peninsula that projected from shore into the anchorage area.

To rectify the situation, Major Fleischmann gave an additional $250,000 for construction of an elbow attaching the west end of the breakwater to the beach at the foot of Point Castillo. This structure, completed in 1930, successfully protected the harbor from further sand deposition from the west—for about three years.

Santa Barbara harbor elbow under construction, October 31, 1929. Black arrow locates Point Castillo; Stearn's Wharf lies beyond the breakwater. The white sand projection in the harbor was formed by waves refracting around the east end. —Fairchild air photo 0-139, Department of Geography, University of California, Los Angeles

Santa Barbara harbor about five years after completion. —Fairchild air photo E-5780, Department of Geography, University of California, Los Angeles

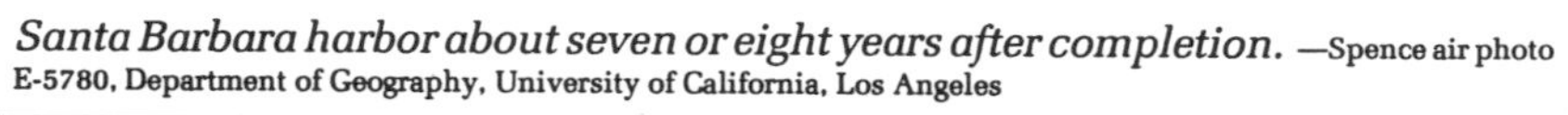

Santa Barbara harbor about seven or eight years after completion. —Spence air photo E-5780, Department of Geography, University of California, Los Angeles

Santa Barbara harbor about nine years after completion. —Fairchild air photo E-2700, Department of Geography, University of California, Los Angeles

Santa Barbara harbor in 1964, 34 years after completion. The sports stadium is visible in the lower left. —Spence air photo E-18971, Department of Geography, University of California, Los Angeles

The primary result, of course, was deposition of a huge quantity of sand immediately west of Point Castillo. West Beach built far out into the ocean, and Leadbetter Beach, farther west, widened. The broad flat now occupied by extensive parking areas, LaPlaya Field, and the sports stadium was deposited in the incredibly short period of less than ten years.

After the area west of the breakwater elbow filled with sand, the longshore drift built a shallow sandbar along the outer side of the breakwater. By 1940, or somewhat earlier, the bar extended to the breakwater's east end. There, refracted swells gleefully seized the drifting sand and dumped it into the harbor, creating shallows and eventually a bar across the harbor entrance. Major storms from the southwest sent their waves to joyfully join in the process of shovelling sand into the harbor.

Meanwhile, East Beach and other prized beaches farther east were starved for sand. They deteriorated rapidly as hungry waves chewed at them, narrowing them and exposing rocks where before there had only been sand. Eventually, this deterioration extended in critical degree as far east as the fine beach at Carpinteria, much to the displeasure of the local populace. According to one report, beaches as far as 45 miles down the coast showed some deterioration because of the Santa Barbara Harbor perturbation. Hotels in Montecito, adjacent to beaches formerly of sand, brought suit against Santa Barbara, which also suffered as waves threatened to undercut Cabrillo Boulevard and a newly constructed pavilion on East Beach.

So, what to do? One obvious action was to clean out the harbor and remove the sandbar blocking its mouth by dredging. Intermittent dredging started in 1935 and continued irregularly through the 1950s. By the early 1960s, a year-round dredging program was initiated, which continues today with federal funding. The dredged sand is carried in a pipe as a water-rich slurry and dumped into the surf east of Stearn's Wharf for redistribution eastward. East Beach has been restored, and the beach at Carpinteria has been partly rebuilt. Beaches farther east along the Rincon are still badly deteriorated. To see a sand-starved, wave-battered, ravished waterfront, visit the eastern part of the Solimar settlement along Pacific Coast Highway on the Rincon between Ventura and Santa Barbara.

Santa Barbara Harbor and other coastal constructions are not the only, or even the major, factor causing deterioration of southern California beaches. A major influence is our flood-control program with its many dams and basins on coastal streams trapping sediment and preventing it from reaching the ocean. Our practice has been to restrict, if not eliminate, floods as dangerous, destructive, and wasteful, particularly in urbanized areas. We do so at the price of serious deterioration of our sandy beaches.

All flat ground in this photo, including parking area, football field, boulevard, and beach (far right) were formed by deposition of longshore sand drift in one decade (1930-1940).

The problem of beach deterioration is currently being attacked by small-scale improvisations. The long-range solution lies in providing greater supplies of sand to the waves devouring the beaches, but a satisfactory source for that sand remains a problem. Recent interest has focused on the possibility of dredging up sand from offshore accumulations and dumping it into the surf for transport to shore.

Santa Barbara Harbor is currently in a state of dynamic quasi-equilibrium thanks to continuous dredging, an expensive process. On problems like this, we need to work with Mother Nature rather than against her. Model studies in experimental tanks could determine what, if any, configuration and arrangement of breakwater structures could provide a protected anchorage and still allow free passage of the longshore drift. It is not clear that such a configuration can be designed, considering the many complex variables involved, but study of the possibility is a challenge worth addressing. If successful, the Santa Barbara breakwater design could become world famous, inasmuch as many other "harbors" face similar problems.

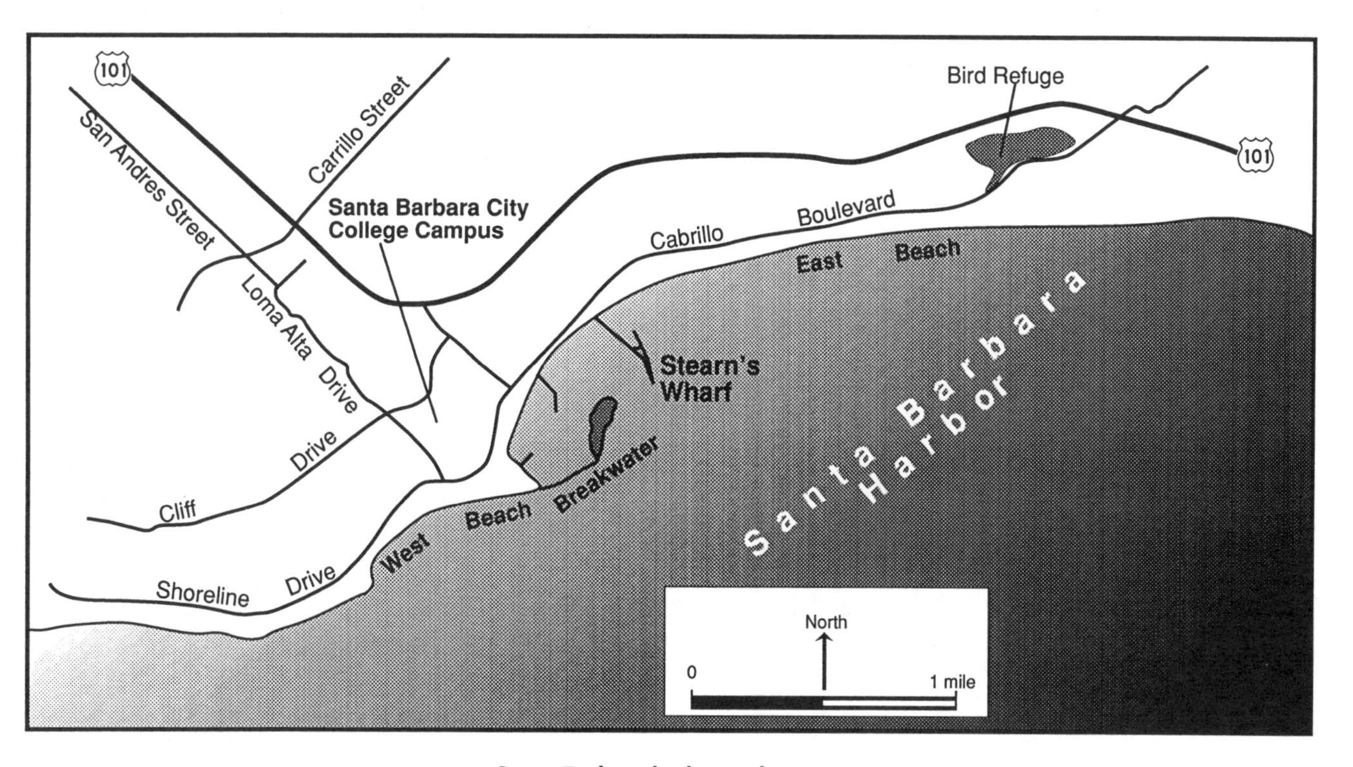

Santa Barbara harbor and environs.

Santa Barbara Harbor and its setting can be viewed nicely from a spot on the Santa Barbara City College campus. The best time is weekends, when school is out. Enter the campus through the kiosk to the south of Cliff Drive. If entering Santa Barbara from the east via U.S. 101, turn off at Cabrillo Boulevard, a left-hand exit, and proceed west on Cabrillo past Stearn's Wharf and Santa Barbara Harbor to Loma Alta Drive. Turn north (right) there and at the top of the hill turn east (right) at Cliff Drive. In a hundred feet or so, turn south (right) at the entrance kiosk of the City College campus.

From the west on U.S. 101, exit at Carrillo Street (not Cabrillo), then go two blocks south (right) on Carrillo to San Andres Street, then east(left) a couple of blocks to a dead-end boulevard stop. There jog a little south (right) to the yellow arrows and ascend the bluff on a gentle grade. This is Loma Alta Drive, but no sign says so. Proceed to the intersection with Cliff Drive and go east (left) about 100 feet to the campus entrance.

If the kiosk is occupied, explain to the attendant that you are studying geological relationships at Santa Barbara Harbor best viewed from Point Castillo at the south tip of the campus. Can you please be admitted and directed where best to park to gain access to that location? If the kiosk is unoccupied, proceed slowly ahead across speed bumps until the campus

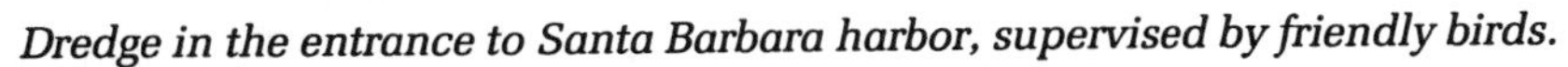

Dredge in the entrance to Santa Barbara harbor, supervised by friendly birds.

opens out and the road curves east. Continue past several low shed-like buildings on the right labelled English, Assessment, and Mathematics until you are south of a spacious expanse of lawn. On the south (right) are vegetable gardens and fruit trees. If possible, park near the first of two "Lifescape—Edible Ornamentals" signs. A trail goes south about fifty feet east of the first sign, along the east side of some banana trees. It leads through the gardens, across grass, past large old cypress trees and trunks to the top of Point Castillo—at the tip of the black vertical arrow of the photograph. It's a walk of only 150 feet. The bluff is a sea cliff whose feet were bathed by waves before completion of the full harbor breakwater in 1930.

By studying the photos, and digesting their captions, you can work out the story of Santa Barbara Harbor for yourself. Remember that what you see all transpired since 1929, and most of it before 1940. It's amazing to realize that the broad flat extending out to West Beach, upon which the sports stadium sits, built up in just a decade solely by accumulation of sand carried by the longshore drift.

Few people deny that the Santa Barbara anchorage is of major commercial and recreational value to the area. It requires much supervision and maintenance and is vulnerable to devastation by storms from the southwest. The search continues for the best solution to operating this valued possession at reasonable cost, with minimum impact on the natural environment.

Many natural systems are like piles of jack-straw; jiggling one factor causes others to shift in unpredictable ways. When humans tamper with natural systems, they often initiate chain reactions that produce undesirable results far beyond expectation. That lesson is clear in Santa Barbara Harbor. We need to proceed gently, carefully, and with as much knowledge and understanding as possible when we tamper with dynamic natural systems, longshore sand drift being just one example.

A walk out on the breakwater is fun, so do it. Go back to Loma Alta Drive, descend to the base of the bluff, and turn east (left) on Cabrillo Boulevard. Take a south (right) turn at Harbor Way, and proceed straight ahead. Except on weekends, parking is usually available on either side among the buildings ahead. Follow your nose and sense of direction to the walkway leading onto the breakwater.

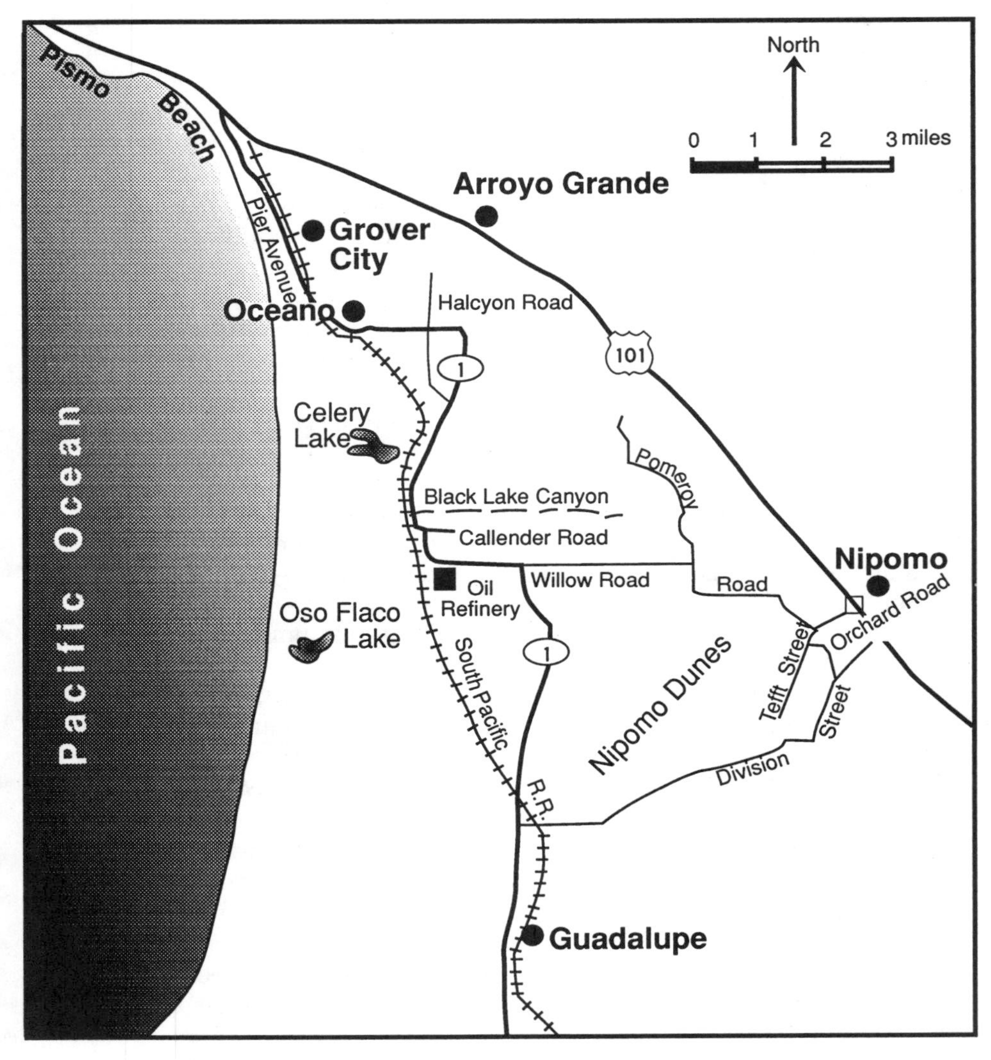

Access map for Nipomo Dunes.

This area, with its many primary and secondary roads, appears on the auto club map of San Luis Obispo County, and on the U. S. Geological Survey, 7.5-minute, Oceano and Nipomo topographic quadrangles. Bring a trowel.

Vignette 6

An Ice Age Sand Lobe

Nipomo Dunes

San Luis Obispo County

Mention sand dunes and many people think of deserts, while others have visions of sea shores. Both are right; dunes favor these two environments. To make a large mass of dunes requires a constantly renewed source of sand and a prevailing effective wind. An "effective wind" blows long enough, many days per year, and hard enough, at least 25 miles per hour, to move sand freely. Ocean beaches are an ideal source of sand, with waves constantly replenishing the supply and no plants to break the wind.

A strong prevailing wind blows across most coasts, typically from sea to land. It carries sand inland, but usually only to where plants break its force, inhibiting sand transport. Wind moves sand off any beach, even wet beaches, where it dries grains on the surface so they no longer stick together. Watch the next time you see a stiff breeze sweeping across a wet beach. Removal of dry, loose sand from any source is duck soup for wind.

Coastal dunes differ from most desert dunes in the large role plant cover plays in determining their location, size, configuration, and eventual stabilization. Plants rarely stabilize most large desert dunes but almost invariably tie down coastal dunes before they are very old.

The California coast boasts at least 27 dune fields of respectable size. Some are narrow belts bordering the inland edge of the beach; others form broad sheets that extend a mile or two inland. A few, such as the large dune field that borders Monterey Bay south of the Salinas River, cover more than 100 square miles and reach five to six miles inland.

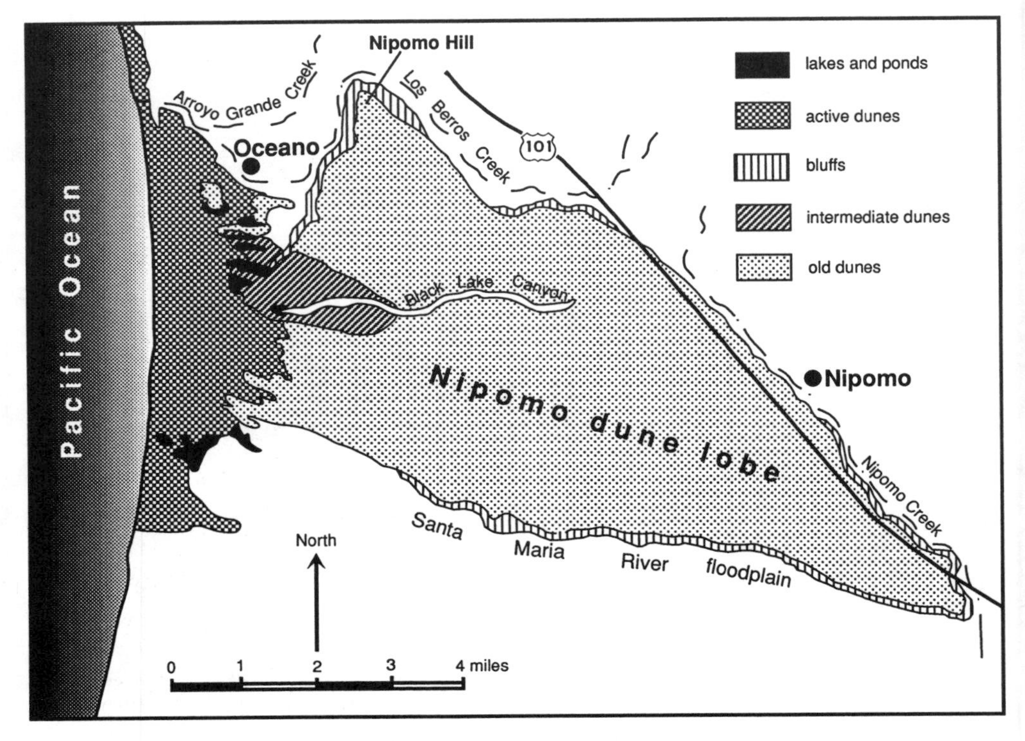

Nipomo Dunes lobe and associated features. —Modified from Wm. S. Cooper, 1967

The broad lowland where the Santa Maria River meets the sea is richly endowed with dunes. Three large tongues of dune sand extend many miles inland. The northernmost and by far the largest, the Nipomo Dunes, is the most spectacular lobe of dune sand along the entire coast of Oregon and California. More than four miles wide, it extends fully 12½ miles inland to a little east of U.S. 101, at Nipomo.

The Santa Maria dune sheets show evidence of different episodes of activity, separated by significant time intervals. When we tell you that this may reflect the behavior of glaciers during the Great Ice Age, your reaction is likely to be, "Well, the old boys blew it this time." We will defend ourselves later.

The Nipomo Dunes can be separated into at least three distinct parts that differ in appearance, and presumably age.

The youngest is a belt of active dunes with no plants growing on them. They appear white on air photos, and extend no more than 1¾ miles inland from the beach. Their dune forms are sharp and fresh, constantly changing. They come to an abrupt and steep inland edge, where they

View to southeast shows active dunes (white) encroaching upon older, stabilized Nipomo Dunes lobe. Note ponded lakes within old dunes near mid-left margin.
—John S. Shelton photo (taken in 1961)

encroach irregularly upon more subdued dunes, most of which lie under a cover of plants, and appear to be older. The advancing active dunes dammed hollows between the older dunes to create a dozen depressions that hold shallow lakes, ponds, and marshes. Fresh ground water keeps them full. It is a bit unusual to find such wet ground in a dune field. Bird watchers find rich and varied bird life along the margins of these waters.

A separate mass of somewhat older dunes surrounds an oil refinery beside the Southern Pacific Railroad, a little south of California 1. These dunes extend about 2½ miles farther inland than the active dunes along the coast. Their sparse plant cover is enough to stabilize them. They appear to be somewhat older than the presently active dunes. These are parabolic dunes shaped like a horseshoe with the open end upwind, a form characteristic of stabilized dunes. Each consists of two nearly parallel sand ridges that join at their downwind ends. The wind blows sand out of the central trough and onto the enclosing ridges as growing plants begin to stabilize the margins of the trough. The parabolic dunes in this part of the Nipomo sand sheet are exceptionally long and sharply pointed, aligned with the prevailing strong west and northwest winds.

Looking west to Celery Lake, one of the water bodies dammed by advancing active dunes, the white band in far mid-distance. View is from the top of the high bluff parallel to California 1.

The oldest and largest part of the Nipomo Dunes, the Nipomo lobe, extends east and southeast from the active dunes along the coast and the stabilized sand sheet around the refinery. These forms hardly resemble dunes. You see a disorganized assemblage of rounded hillocks and hollows, but here and there are exposures of homogeneous sand with the uniform grain size characteristic of dunes. At most, the hills stand about 80 feet above the hollows. The degraded shapes of these dunes suggest that they have lain beneath plant cover for many years. Iron oxide locally cements the dune surfaces into a crust and stains the sand a dark brown color. The plant cover is mostly the scrubby chaparral brush typical of so much of southern California, with a few struggling oak trees.

Black Lake Canyon cuts a deep gully through the middle of the oldest sand sheet from east to west for 3½ miles. It ends in Black Lake. The canyon, as much as 200 feet deep, must have been eroded after plants stabilized the sand. The depth of the canyon also suggests that this largest part of the dune tract is fairly old. These older dunes cover a triangular area over four miles wide in places, projecting eleven miles inland from the active dunes. The tract seems to have covered a relatively smooth surface that rose to an upland flat, now so dissected that it stands as much as 300 feet above adjoining terrain.

Steep, stream-cut banks up to 150 feet high border the Nipomo lobe on its north, south, west, and northwest margins. Near Nipomo Hill, at the lobe's northwest corner, the banks are more than 300 feet high where they overlook the truck gardens in the valley of Arroyo Grande Creek. But much of that high face is eroded into the Paso Robles formation, a deposit of stream sediment that lies beneath some 50 to 100 feet of dune sand. The present size and shape of the oldest part of the Nipomo sand sheet are a result of erosion and do not necessarily reflect the original extent of the dune tract. This erosion of its margins is another indicator of a ripe age for the Nipomo lobe.

Much of the oldest sand sheet is now urbanized. Roads, houses, ranches, excavations, lemon and avocado groves, and replacement of original vegetation with imported trees, including large plantations of eucalyptus, have drastically modified the original surface. But about 3,000 acres of the old Nipomo sand sheet are now protected, through the efforts of the San Luis Obispo chapter of The Nature Conservancy.

How about those ice-age glaciers that never reached anywhere near this coast? What did they contribute to the formation and history of the Nipomo Dunes? How could they possibly contribute anything?

If all the world's present glaciers were to melt, sea level would rise more than 200 feet. When the enormous ice-age glaciers formed at various times during the last million years, they tied up enough water to drop sea level as much as 450 feet, depending upon which ice age was involved. When ice-age glaciers were near their maximum, and sea level was at its lowest, the shoreline here lay about eight or nine miles seaward of the present beach. Dunes probably formed along that shore—but they were doomed. The rising sea level submerged them when the glaciers melted at the end of the ice age.

Old dune masses well above present sea level, such as the Nipomo lobe and thoroughly stabilized dunes around the refinery, are thought to be relics of former high stands of sea level. The old Nipomo lobe probably formed between ice ages when sea level was at least as high as it is now, perhaps higher. Sea level is now high compared to the levels of the past 10,000 years, and the active part of the Nipomo sheet is the result. To make the much larger older part of the dune sheet would require a longer interval of high sea level, perhaps more sand than the modern beach supplies, and possibly stronger winds than those that now blow. What we know of the record of the great ice ages and the intervals between them suggests that such conditions may have existed about 160,000 years ago. Perhaps that was when the wind blew the oldest part of the Nipomo dune field inland from the beach.

An excellent network of roads provides access to the dune fields. The best way to reach them is to leave U.S. 101 at Nipomo and drive west on Tefft Street about 0.7 mile to Pomeroy Road. Turn northwest (right), and drive about 0.2 mile along the northeast side of Nipomo Regional County Park. About 200 feet beyond the last baseball diamond, turn southwest (left) into the park on a narrow paved drive, and continue a short way to a large parking area on the south (left). Walk 50 feet up the road to a sandy trail marked by a gas pipeline sign. Climb 100 yards of sand to the top of a rounded knob to view the knob-and-hollow topography of the degraded older dunes and look at their cover of chaparral scrub and small oak trees.

Then continue on this drive to Tefft Street, turn north (left) a short distance, then turn east (right) onto Orchard Road. You will be driving across the top of the Nipomo lobe. Watch for fresh excavations and cut banks. They show a foot or two of dark brown sand at the top, locally cemented well enough by iron oxide to break into angular clods. About three feet down, as the abundance of iron oxide decreases, this dark sand gradually gives way to yellow sand, which becomes nearly white at a

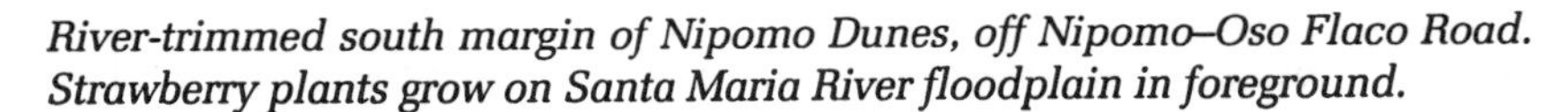
River-trimmed south margin of Nipomo Dunes, off Nipomo–Oso Flaco Road. Strawberry plants grow on Santa Maria River floodplain in foreground.

depth of five to seven feet. Many long years of weathering and decomposition of iron-bearing minerals were necessary to create that gradation in color.

Continue on Orchard Road to the intersection with Division Street, turn south (right) and drive to the brink of a bluff about 130 feet high. This is the northwest edge of the Nipomo lobe where it overlooks the Santa Maria River valley. You look directly down on the truck gardens on the valley floor and west off in the distance to the white inland edge of the advancing dunes along the coast.

Follow the road down the bluff toward the river. Nothing you see suggests that this deep river cut exposes anything but dune sand, but of course it is possible that a veneer of sand sliding down from above could cover other rocks. An apparently reliable geologic map, however, suggests that in places the sand of Nipomo lobe may be more than 200 feet thick. Layers of dark brown sand exposed in small gullies in this bluff suggest that the river may have cut through a stack of sand layers of different ages.

Natural chaparral vegetation complex on the older part of Nipomo dunes, viewed here from the intersection of Willow and Pomeroy roads.

At the base of the bluff, turn around and go back to Pomeroy Road, then turn northwest (left). The Nipomo Dunes are well settled. Most of the natural vegetative cover has been removed, except for oak trees; the pine, cypress, walnut, and eucalyptus trees have been planted. In about three quarters of a mile, Pomeroy Road straightens and heads directly west for nearly a mile. Watch for linear dune ridges on both sides of the road, the long sides of parabolic dunes.

After Pomeroy Road curves north, turn west onto Willow Road and note that the area immediately to the north retains a thick cover of the original oak trees and chaparral scrub. Willow Road heads west one mile to California 1, which crosses the southern part of the dune sheet. Continue straight ahead on California 1 across sand that is much paler than that near Nipomo. This sand is much younger and has not weathered enough to acquire a brown iron oxide stain. From near Willow Road Industrial Park, you can see the towers of the oil refinery to the south. The dominant native vegetation cover of grass and scattered bushes on these dunes is sparser and distinctly different from the oak and chaparral cover farther inland. These are the dunes of intermediate age.

As you continue west on California 1, near the entrance to the oil refinery, you see the steep front of the active dunes advancing onto the dunes of intermediate age. California 1 turns north, meets Callender Road, where it jogs west, then dips to cross lushly vegetated Black Lake Canyon. A short distance north of the canyon, it runs near the brink of a high bluff past a splendid view of the lakes, ponds, and marshes dammed within hollows in the older dunes. You get a clear view of the inland edge of the active dunes advancing onto the floor of Cienega Valley by turning onto Halcyon Road and driving down the bluff.

For a closer look at the active dunes and the beach that supplies their sand, turn west (left) onto California 1 at Halcyon Road, and drive through Oceano to Pier Avenue, follow it to its end at Pismo State Beach, Oceano Campground, and Pismo Dunes Vehicular Recreation Area.

If the tide is out you can go clamming on Pismo Beach.

VIGNETTE 7

ROOTS OF ANCIENT VOLCANOES

Morro Rock and Other Knobs

SAN LUIS OBISPO COUNTY

Everyone who visits Morro Bay remembers the view of Morro Rock, an isolated knob rising steeply to a rounded top 580 feet above the ocean. This high dome looks so unusual that we wonder what it is and how it formed. Was it pushed up from below by some jolly subterranean giant just to befuddle people? Or is it simply a mass of hard and tough rock that resisted weathering and erosion while less durable rock on all sides succumbed? Can anyone explain its shape, setting, and origin? Let's examine the evidence.

To begin, drive along the shore out to Morro Rock for a close look. Large blocks of rock, some with fresh fracture faces, good for mineralogical inspection, lie at the foot of the cliff to the west (left). Do not mistake the small circular patches of pale gray lichen growing on weathered rock surfaces for mineral grains.

Fresh surfaces show that about half the rock consists of a fine-grained, gray, homogeneous groundmass. Visible crystals constitute the rest of the rock. Scattered through it, you see abundant larger (as much as a half inch long) crystals of a sparkling, light-colored mineral, and a good many smaller, light and dark mineral grains. The recognizable, well-developed crystal forms and generally massive appearance indicate that this is an igneous rock. The large size of the crystals and the shape of Morro Rock and other similar knobs suggest that it is probably an intrusive rock, and the fine-grained matrix indicates that it probably crystallized at very shallow depth.

Oblique air view of Morro Rock from the northwest, with Morro Bay in midground, dunes on bay bar and to left of Morro Rock. Photo was taken in 1947. —Spence air photo E-13316, Geography Department, University of California, Los Angeles

The rock is hard and coherent, which helps it resist erosion. Many surfaces display banding that probably formed as the molten rock flowed and intruded into the surrounding host rocks. Several different sets of parallel fractures, called joints, break the rock into regular blocks. For geologists, a joint is not a place to get a beer, but rather a fracture plane passing through otherwise massive rock. Blocks breaking away along joints created the craggy irregularities in the steep sides of Morro Rock.

Morro Rock's circular outline expresses the shape of the igneous intrusion. Were the rock into which it intruded still there, you might think the igneous mass looked like a plug. That is a good name for it: an igneous plug or neck. The rounded top is largely the result of erosion—partly a matter of slabs of rock flaking off the intrusion and partly the blunting of sharp corners as weathering decomposes and disintegrates the rock.

Is there more to the story than this one knob reveals? It is not convenient to explore northwestward onto the sea floor, so let us head southeast along California 1, keeping an eye peeled for other versions of Morro Rock. Black Hill, a circular, round-topped conical knob 660 feet

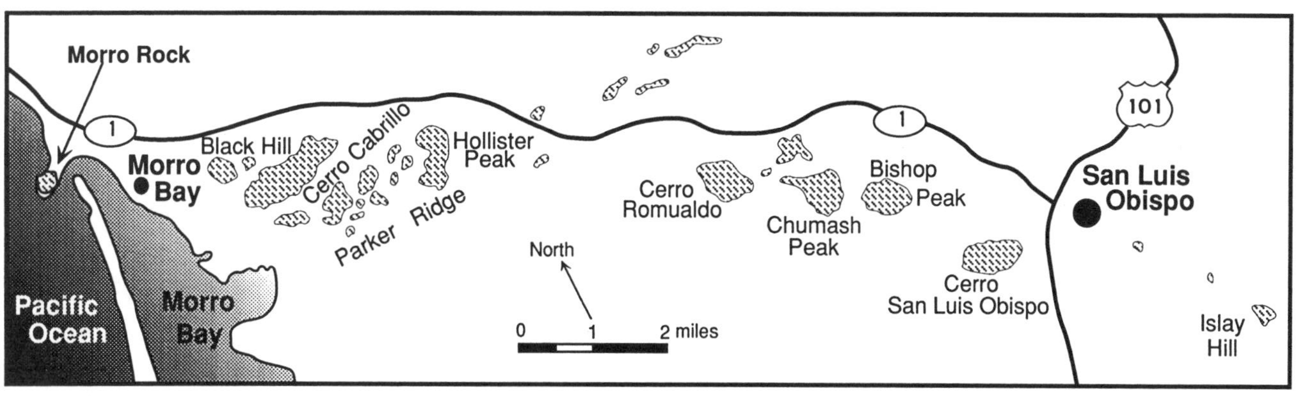

Morro Rock and chain of knobs underlain by intrusive rocks (stippled pattern).

For the easiest route to Morro Rock, take the Main Street exit off California 1 into the city of Morro Bay. Follow Main Street west into town as far as Beach Street, then drive north on Beach and downhill to Front Street. Continue north on Front past the power plant with three towering stacks. Proceed ahead on the obvious road to Morro Rock and turn north into the large dirt parking area on its inland side.

high, rises about one-half mile south of the highway, just northwest of the off-ramp to Los Osos and Baywood Park. A half mile farther, Cerro Cabrillo rises south of the road, an elongate narrow ridge 910 feet high. Another mile brings you opposite Hollister Peak, 1,410 feet above sea level with an irregular ragged crest. It lies at the east end of a large hilly area called Parker Ridge.

Some geological maps show Parker Ridge as a single intrusion with an irregular shape; more detailed maps show that it is actually a cluster of many small intrusions that may be the partly exposed top of a single large intrusion. More recent mapping also identified some small intrusions northeast of California 1 and a few outcrops of associated lava flows.

Southeast of Parker Ridge, you pass through nondescript low hills for about three and one-half miles. At the entrance to Cuesta College, the first two of four successive knobs rising to 1,546 feet above sea level appear ahead and to the right. In order, these southeasterly aligned knobs are Cerro Romualdo, Chumash Peak, Bishop Peak, and Cerro San Luis Obispo. This alignment extends into San Luis Obispo, where Cerro San Luis Obispo lies just one-half mile west of the 101 Freeway, about midway between the Marsh Street and California 1 exits. People coming into San Luis Obispo from the south on 101 see this high knob ahead and to the north (left), between the Madonna Road and Marsh Street exits.

Geological maps show at least thirteen separate igneous intrusions or clusters of small intrusions within a southeast-trending zone that extends nearly eighteen miles from Morro Rock to Islay Hill, about three miles southeast of San Luis Obispo. This chain of plugs was first mapped in 1904 by a fine early American geologist, Harold W. Fairbanks (1860-1952). The intrusions lie along a gently curving arc bowed to the northeast.

If you were to inspect them closely, you would find that all these knobs are composed of rocks like that of Morro Rock. Although they differ slightly, their chemical compositions are so similar that these must be closely related igneous intrusions. Their arrangement suggests that a zone of more or less parallel fissures brought molten magma nearly to the surface. Those fissures are now dikes, fractures filled with igneous rock. The distribution of the intrusions makes it seem likely that they are aligned along a linear zone of short and overlapping parallel dikes rather than on a single large dike. But the shortage of definitive exposures causes that conclusion to depend more on speculation than direct observation.

Why should dikes change upward from thin sheets of igneous rock filling fractures to nearly circular plugs or irregular intrusions? Perhaps the high pressure at depth helps the host rock to maintain planar fractures, whereas wall rock closer to the surface is likely to be more completely shattered, permitting intruding magma to carve out chambers of various sizes and shapes. At Kilauea Volcano, on the island of

Hawaii, many eruptions begin as linear "curtains of fire" as lava erupts from a long fissure. Subsequently, the activity focuses into a centered eruption as magma crystallizing within the fissure concentrates the flow into a narrow pipe. Something similar may have happened between Morro Rock and San Luis Obispo.

The fine groundmass of rocks in our line of intrusions suggests that they lay close to the surface, where they could cool fairly rapidly. Most host rock at shallow depth is saturated with water. If rising molten magma suddenly comes into contact with this water, a steam explosion commonly follows. The explosion may blow a hole through to the surface, forming a crater and causing broken volcanic rock to accumulate in and around it. Molten rock may erupt through the hole, forming lava flows.

It seems likely that a line of active volcanoes once loomed over the San Luis Obispo area, spewing volcanic rubble across the countryside, filling the sky with volcanic ash, and erupting lava flows. Sedimentary formations in this area contain volcanic debris that could have come from this old volcanic field, which was active about 25 million years ago. Erosion has since destroyed much of the evidence, leaving us with a string of igneous intrusions, the roots of the volcanoes, that withstood erosive processes. Several of the intrusions now stand 1,000 feet above the surrounding terrain of softer host rocks. The imposing edifice of Morro Rock owes its present form to the erosional forces that stripped away the surrounding softer rocks and left the harder volcanic plug behind.

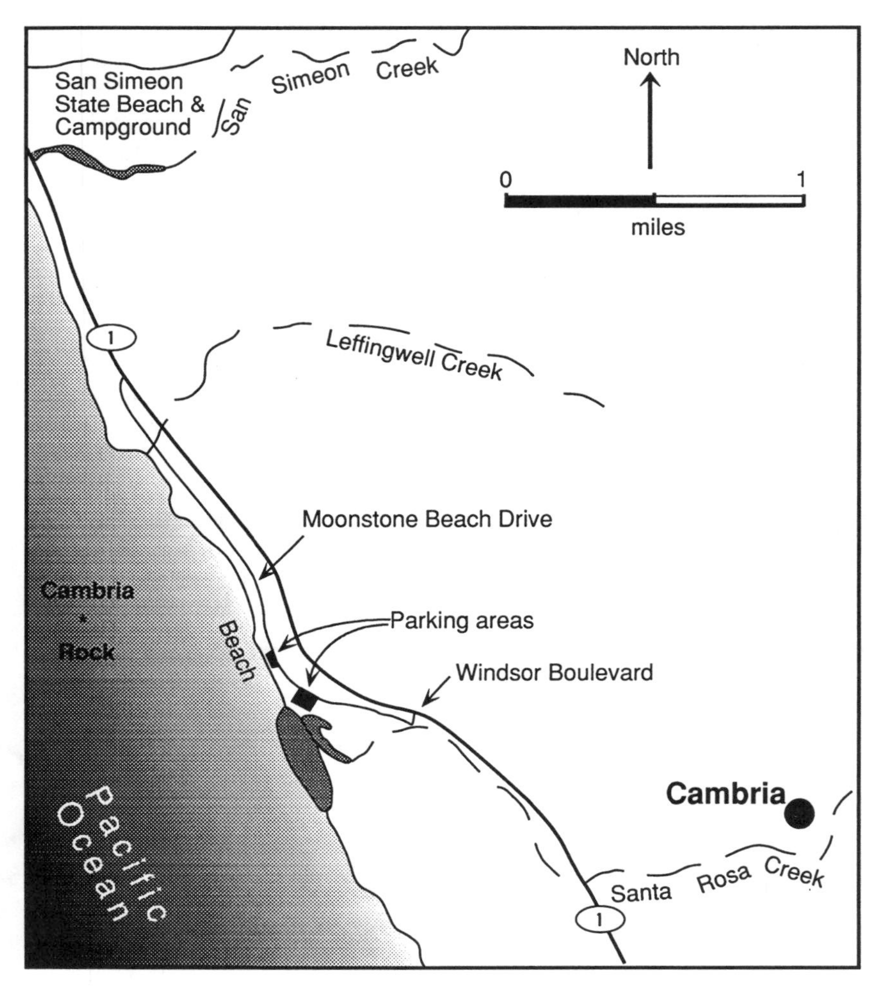

Cambria segment of San Simeon coast.

For some of the best pebble-hunting sites, turn off California 1 in Cambria at Windsor Boulevard. Jog seaward a bit and go northwest about 0.4 mile on Moonstone Beach Drive to a large, paved state beach parking area. Access to the beach here is easy by way of several walked-in trails. Once on the beach, go northwest to where the waterline is close to the sea cliff.

VIGNETTE 8

SAMPLING THE MYSTERIOUS FRANCISCAN FORMATION

San Simeon Beach Pebbles

SAN LUIS OBISPO COUNTY

Pebble hunting is a bit like fishing—some days are better than others, but a day spent admiring pebbles on the San Simeon coast surely doesn't count against your longevity. Just about everyone—youngsters and oldsters, girls and boys, doctors and lawyers, market speculators, sports fans, and beach bums—love looking at the smooth, lustrous pebbles found on many mixed sand and pebble beaches. All you need is a discerning eye, protection from a cool sea breeze, and back and legs flexible enough to allow you to pick up stones for a closer look. But beware: Inspecting pebbles can be addictive.

The San Simeon coast is famous for the beauty and variety of its beach pebbles, with the best beaches to comb lying between Cambria and the San Simeon State Beach campground at San Simeon Creek. Most high tides, and especially those associated with storms, rework the beaches, so conditions change constantly. A good time to look for pebbles is after a major storm. Low tide is obviously better than high tide.

The favored beaches are mostly narrow with pebbles usually most abundant in small coves or near the mouths of streams that bring stones to the shore. A 25- to 50-foot-high sea cliff blocks easy access to most of the beaches, but you can generally find a passable trail down the bluff.

On Moonstone Beach, that stretch of San Simeon State Beach extending from 0.4 mile north of Windsor Boulevard to Leffingwell Creek, look for well-worn, smooth, and beautifully lustrous pebbles one-quarter to one inch across. They lie clustered around beached tree trunks, large rocks, and irregular projections of the cliff—obstructions that created centers of unusually high turbulence in the wave swash. The waves carried away the sand, leaving concentrations of pebbles.

Here and elsewhere along Moonstone Beach, most pebbles are very finely crystalline rocks, both dense and hard, mostly chert. Chert is a form of silicon dioxide, silica for short. Silicon and oxygen are, by weight, the two most abundant elements in the earth's crust. Silica is the building block of the common minerals quartz and chert. Rocks consisting solely of chert also go by that name. Quartz forms beautiful crystals; some of the larger and clearer ones adorn bracelets and necklaces. The silica crystals in chert are so small they are indistinguishable even under microscopic magnification, hence the term cryptocrystalline (crypto means hidden)

A typical setting for good pebble hunting: sea cliff and pebble accumulations on Moonstone Beach near Cambria.

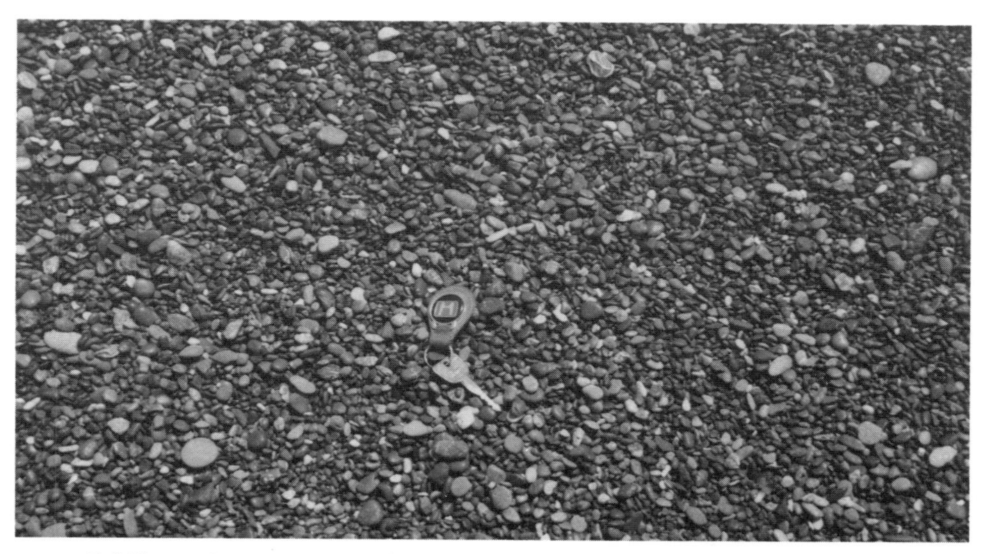

Pebbles on Moonstone Beach near Cambria. (Key and fob are 5 inches long.)

to describe its texture. The closely intergrown fibrous crystal structure of chert gives it a distinctive homogeneity, a somewhat waxy luster, and a smoothly curved, shell-like (conchoidal) fracture habit.

Chert's toughness and homogeneity have made it a useful and valuable commodity throughout human history. Native Americans fashioned arrowheads and spear points from chert. They prized the chert for its hardness, durability, attractive colors, and, most of all, its fracture habit, which permitted them to chip sharp edges. The ore grinders at many early mining operations were ball mills that used chunks of chert as the grinding balls. In modern times, manufactured wafers of silica form the cornerstone of the transistor and computer industry.

Chert pebbles on Moonstone Beach come in a wide range of colors but are mainly reddish brown and green. You can assemble an amazing spectrum of shades of green, ranging from clear apple green through a variety of grayed greens to some very dark greens. Other colors include light and dark gray, pale brown, white, and an occasional clear red. In addition to the chert, you will find a few pebbles of volcanic and other igneous rocks, as well as a few chips of sedimentary rocks. None of those are as smooth and lustrous as the pebbles of chert, and most contain at least a few visible mineral grains.

A very few of the green San Simeon pebbles are jade, a hard, green, glassy mineral called jadeite. Jadeite is one of the most physically resistant of all minerals and extremely difficult to break, so it survives the grinding action of a beach environment very well. You'll find most jadeite a dull green that would hardly make a gem. A few specimens are bright and translucent enough to deserve polishing. But it has been our sad experience that most of the small, green San Simeon pebbles that looked like jade proved, upon being broken, to be green chert. If they break easily, you can be sure they are chert.

A little farther northwest on Moonstone Beach is also a good place to look at pebbles. A block beyond Weymouth Street, wooden steps lead to the beach. Parking is available, except on crowded weekends, all along the top of the sea cliff here, with additional parking about a block farther northwest. You will find the best pebble accumulations on this beach 100 to 300 yards southeast of the steps, where shoreline rocks and an irregular cliff create turbulence in high-tide surfs.

For somewhat larger pebbles, many of reddish-brown chert, go to the beach below the parking area at the vista point just south of San Simeon Creek. These stones are mostly ½ to 1½ inches in diameter, rounded, and smooth. Some are bits of rather coarsely crystalline rocks and not as smooth and lustrous as the chert pebbles on Moonstone Beach. Careful searching, however, will always turn up a few nicely smoothed samples of dense chert.

Still larger pebbles, many big enough to qualify as cobbles (larger than 2½ inches), lie on the beach at Pico Creek, 2.5 miles north of San Simeon Beach Campground. This spot is accessible via a parking area and a wooden stairway. Watch for the beach-access sign along the highway near the north end of the cluster of motels and restaurants south of Pico Creek. Scattered among the larger, mostly coarsely textured but well-worn and nicely shaped stones are a few smaller, lustrous pebbles of fine, dense chert. North from Pico Creek to San Simeon Village, the beaches are almost entirely sand, poor places for pebble-hunting.

Why are San Simeon pebbles so delightfully varied in color, texture, and luster? It is because they come from an unusual and complex bedrock source. They are not the average run of beach stones, and their exceptional beauty and character reflect the unusual origin of their parent rocks. Most of these pebbles come from the Franciscan rock assemblage. It underlies large areas in the hills inland from San Simeon coast. You can see it in the sea cliffs north from San Simeon Creek.

The Franciscan assemblage has for many years been one of the major mysteries of California geology. Like most rock units, it was named after a geographic locality, in this case the San Francisco Bay area. The assemblage is more widely exposed in northern and central California than in the southern part of the state. The pebbles tell us that there is a lot of chert in the Franciscan assemblage.

All silica is hard and resistant—physically and chemically—because its silicon and oxygen atoms are held together by exceptionally strong electron bonding. As a result, substances made of silica, such as quartz and chert, are survivors within the geological milieu. For example, the beautiful white sand beaches of the western Florida coast around Pensacola are almost wholly quartz grains that have survived hundreds of millions of years of weathering and physical battering before arriving on the beach. Essentially all other minerals have deteriorated, leaving only the quartz. Being just about as tough as quartz, the cherty San Simeon pebbles are also survivors.

Native Americans in the West found most of their chert as nodules or thin regular layers enclosed within limestone beds widely exposed in our western mountains. We know that calcium carbonate ($CaCO_3$), the building block of limestone, accumulates on the seafloor, so the chert must have originated there too. Sea water contains considerable dissolved silica. In some special situations, the silica precipitates chemically, but more commonly, simple single-celled plants (diatoms) or marine protozoa (Radiolaria) use the dissolved silica to build siliceous skeletons. When diatoms and Radiolaria die, their skeletons drop to the seafloor where they accumulate in huge numbers as opaline siliceous

A representation of how trough sediments, deep-sea deposits, sea floor crust, and upper mantle rocks could be scraped off the down-plunging Pacific plate and plastered onto the overriding North American plate to form the Franciscan assemblage.

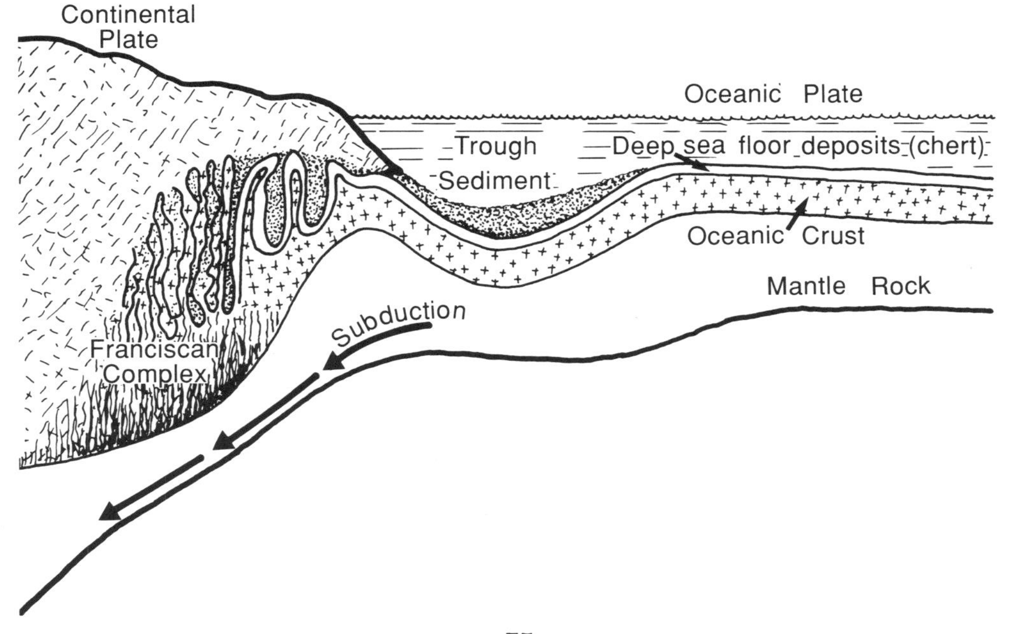

ooze. This substance then goes through several amorphous and crystalline stages before becoming chert.

The cherts in the Franciscan assemblage occur mainly as thin regular layers called bedded cherts. The origin of bedded cherts was something of a mystery until extensive drilling and coring of the deep-sea floor showed that bedded cherts are common there. They build up on the deep-sea floor, so deep that physical and chemical conditions prevent the accumulation of calcium carbonate. These cherts contained traces of the skeletons of diatoms and Radiolaria plus siliceous sponge spicules (needlelike rods that serve as part of the sponge's skeleton), establishing those organisms as the precipitators of the silica.

The question now confronting us is how did deep-sea cherts come to be part of the Franciscan assemblage? Before the San Andreas fault developed—less than 100 million years ago but more than 30 million years ago—the North American plate overrode rather than slipped past the Pacific plate. Small land masses (called microcontinents), deep-sea sediments, and parts of the underlying oceanic crust were scraped off the top of the down-plunging Pacific plate and plastered onto the edge of the North American plate. These accreted geologic fragments make up the Franciscan assemblage.

After the Franciscan cherts became part of North America, they were lifted within mountains that rose near the coast. Erosion ultimately exposed them on the surface and eventually broke them into pieces that runoff and streams carried down to the seashore. There, the waves tumbled the stones about, leaving behind the polished pebbles that now cover the beach. If these pebbles could talk, what a yarn they could spin.

Vignette 9

A Boon To Communication

The Whittier Narrows

Los Angeles County

Drivers from Duarte hustle south on Interstate 605 for a day at the beach, probably without wondering how they passed through the Puente Hills without crossing over any hills. The Puente Hills form a formidable barrier that extends east and southeast from Elysian Park to the Chino Hills, separating the San Gabriel Valley from the Los Angeles basin. Whittier Narrows is the only gap in this thousand-foot-high hilly barrier. It appears on Auto club maps of greater Los Angeles and the U.S. Geological Survey, 7.5-minute, El Monte topographic quadrangle map.

The San Gabriel River cut that gap, two miles wide and 800 feet deep. Whittier Narrows slices the Puente Hills into two parts: the Repetto Hills to the west, and the Whittier and Chino hills to the east. Interstate 605 passes through Whittier Narrows, along with many other highways and streets, pipelines, powerlines, flood-control channels, and other lines of communication. How did the San Gabriel River breach the Puente Hills?

The answer is somewhat unusual, by classical geological standards: The river was here before the Puente Hills rose. That is likely to happen only in regions where the earth's crust is actively deforming. Some of the rocks folded in the Puente Hills are so young that it seems likely the hills are still rising. Indeed, the earthquakes that originated deep beneath Whittier Narrows in 1988 and 1989 show that the region is under stress today.

Before the Puente Hills rose, the San Gabriel River and several smaller streams flowing south out of the San Gabriel Mountains had established shallow channels across the area. Then a low ridge, the first inkling of the

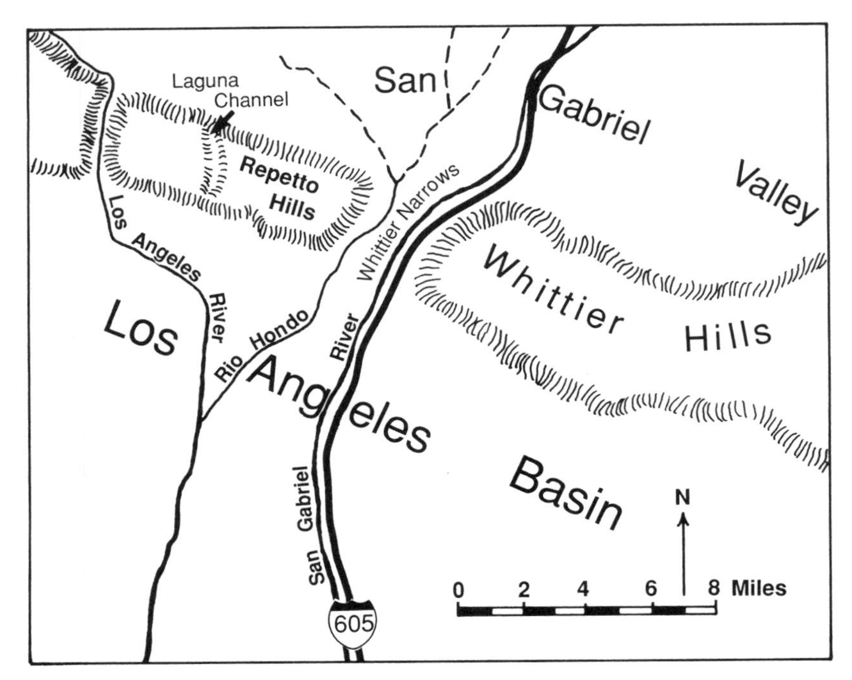

Diagramatic representation of Puente Hills uplift separating the Los Angeles basin and San Gabriel Valley, and antecedent Whittier Narrows.

future Puente Hills, started to rise across their courses. Some of the smallest streams were probably turned aside at once. But the larger, more powerful streams, including the San Gabriel River, were able to erode their beds as rapidly as the ridge rose, cutting ever deeper channels across the growing barrier.

In due time, one after another of these streams was unable to challenge the rising ridge, and was shunted aside. Some of these defeated streams turned east to become tributaries of the San Gabriel River, augmenting its flow and making it more able to erode its channel. Finally, only the San Gabriel River survived, still breaching the barrier of the Puente Hills. The hills continue to rise, and the river continues to cut its way through as it flows south to the ocean.

Valleys that carry streams through mountain ridges are usually called water gaps. Whittier Narrows is a water gap. As the rising ridge raised the abandoned stream valleys, it converted them to wind gaps. Water gaps and wind gaps are not unique to southern California. They abound in other parts of our country, especially the Appalachian Mountains, where Cumberland Gap is a good example. But most water gaps elsewhere were created as streams cut their way down through masses of hard rock,

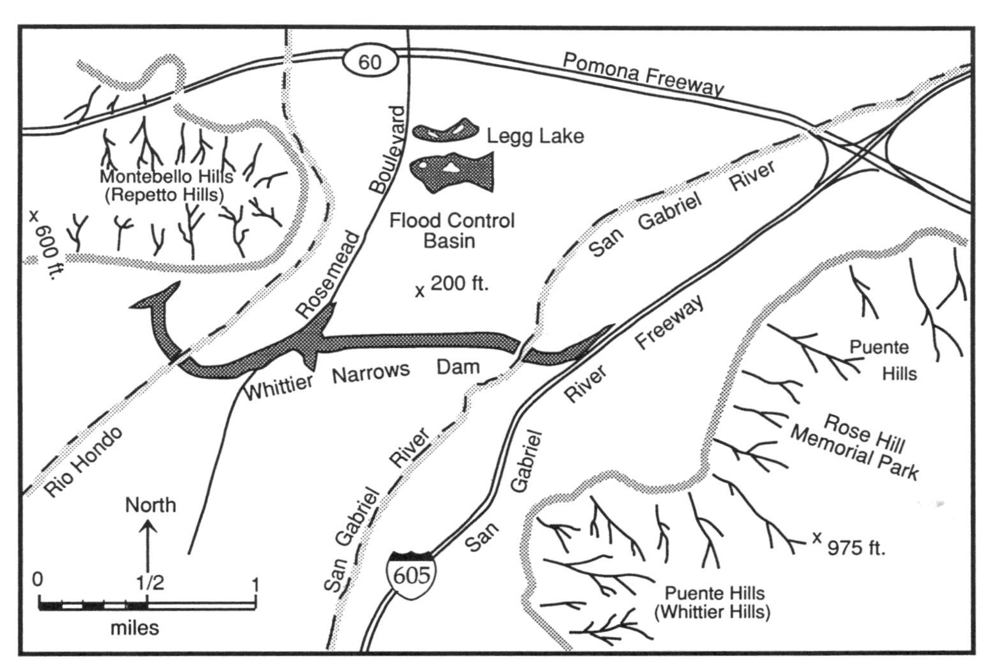

Whittier Narrows and environs.

which subsequently became ridges, as the surrounding softer rocks were eroded.

On clear days, people looking south from tall buildings in Pasadena can see the wind gaps indenting the crest of the Repetto Hills, west of Whittier Narrows. Four of these gaps provide important routes of communication. From the west those are Laguna channel, the route of the Long Beach Freeway; Coyote Pass, traversed by Monterey Pass Road; the wide channel that Atlantic Boulevard follows; and a little farther east, another unnamed pass used by Garfield Avenue. Although the smaller streams that cut those passes finally lost their battles against the rising Puente Hills, we can thank them for their partly successful fight. Their wind gaps make it easier to pass over the Puente Hills, and to understand what happened as they rose.

Rio Hondo is a curious minor anomaly within Whittier Narrows. It is a wet-weather stream with at least two branches that drain parts of Pasadena-Alhambra and Monrovia-Arcadia. Rio Hondo follows a course through Whittier Narrows a mile or so west of the San Gabriel River. Upon emerging from the south end of the narrows, it turned southwest to join the Los Angeles River between Downey and South Gate, and its artificially confined flood control channel still follows that route. Why Rio Hondo preferred this roundabout course to the ocean by way of the Los

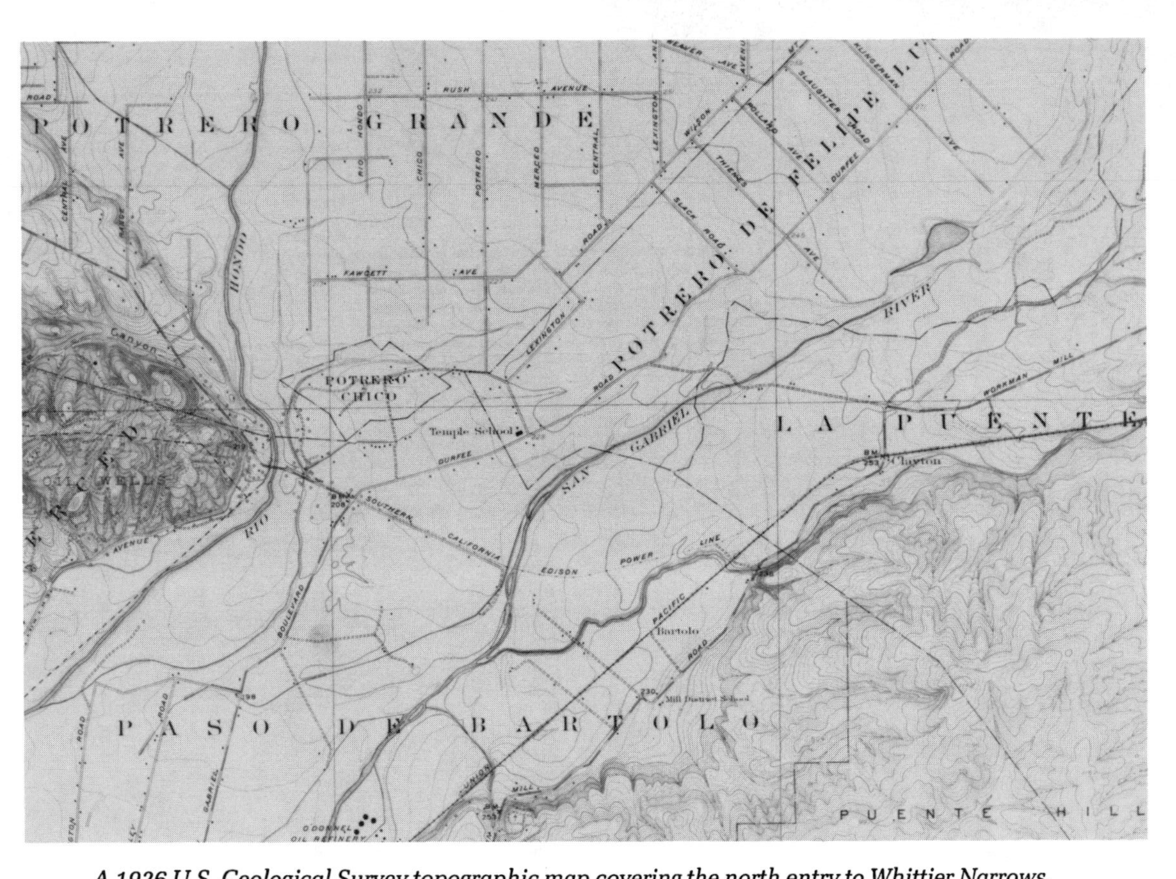

A 1926 U.S. Geological Survey topographic map covering the north entry to Whittier Narrows, with Repetto Hills (5-foot contours) on the left and Whittier Hills (25-foot contours) on the lower right. Separate channels of San Gabriel River and Rio Hondo shown.

View in 1967 east across north entry to Whittier Narrows. Interchange of north-south San Gabriel River freeway (Interstate 605) and east-west Pomona freeway at mid-left. The confined San Gabriel River bed extends obliquely to lower right corner; Whittier Hills are in the upper right. Compare with the 1926 topographic map on page 80. —Spence air photo E58-164, Geography Department, University of California, Los Angeles

Angeles River is a mystery. Perhaps the large amount of sand and gravel carried by the San Gabriel River raised its channel just enough that Rio Hondo would have had to flow uphill to join it.

The next time you whisk through Whittier Narrows on the Interstate 605, tip your hat to San Gabriel River and its hard-won victory in the long battle with the persistently rising ridge of the Puente Hills. Rivers do most of their work of erosion, transport, and deposition during floods. Huge gravel deposits, rich in cobbles and boulders, west and south of Azusa and the erosion that was needed to cut Whittier Narrows testify to the river's former power. The river's floods must have been holy terrors before dams and huge flood impoundments robbed the river of its strength. San Gabriel River deserves a bronze plaque somewhere in Whittier Narrows, celebrating its contributions to the people of the greater Los Angeles area.

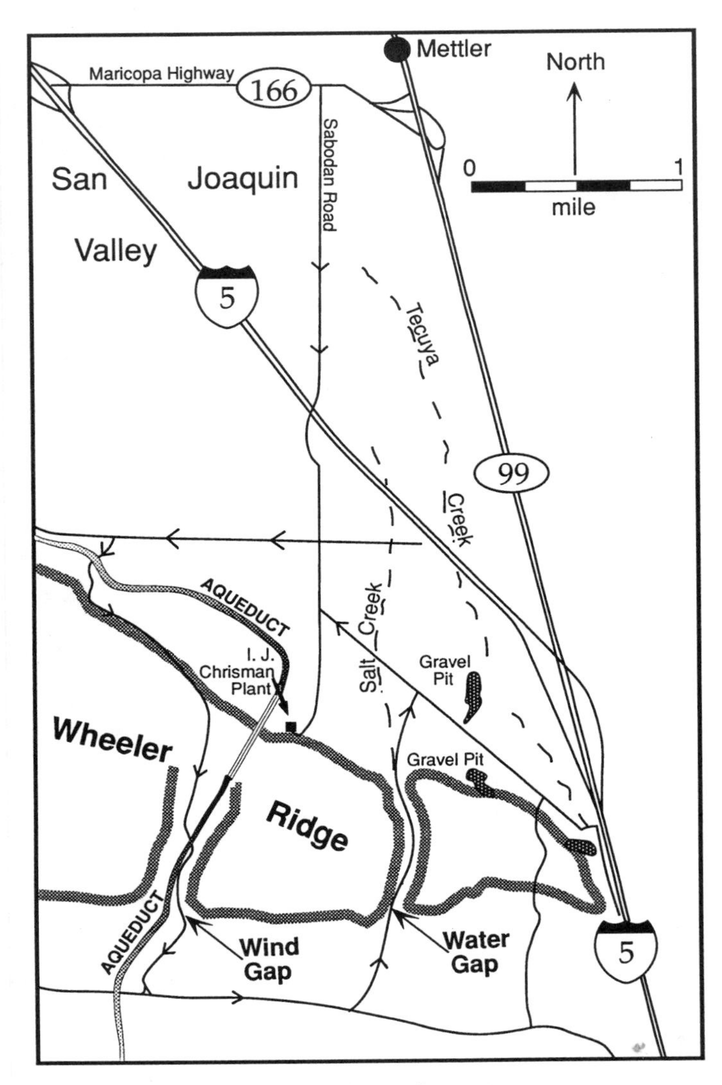

Wheeler Ridge and vicinity.

To get a close look at Wheeler Ridge, exit Interstate 5 at California 166, go east on California 166 (Maricopa Highway) a little more than one mile and then south on Sabodan Street. Seven-tenths of a mile after crossing Interstate 5 on an overpass, turn west (right) on an unnamed, wide, paved street. Continue for one mile, then turn on the first paved road to the south (left), headed for the bridge crossing the aqueduct. Keep east (left) on the paved road, ascending to the wind gap and park on the near side of the canal. If traveling on California 99, exit at California 166, go west to Sabodan Street, and follow the instructions above.

VIGNETTE 10

WHEELER RIDGE

Its Gift to the Edmund G. Brown (California) Aqueduct

KERN COUNTY

Traffic is heavy in and out of the San Joaquin Valley along the Golden State Freeway (Interstate 5) and California 99 near Grapevine. The junction of these two major highways lies just north of the east end of Wheeler Ridge, so thousands of people pass the ridge every day. You can also see it on the auto club Kern County map and the U.S. Geological Survey, 7.5-minute, Mettler topographic quadrangle.

Views are best for southbound travelers on Interstate 5 and reasonably good on southbound California 99. A little careful sidewise looking by northbound motorists on either route reveals the essential features. In either direction, motorists should have no trouble spotting the four large pipes rising from a pumping plant on the Edmund G. Brown, or California, aqueduct on the north side and near the east end of Wheeler Ridge. They cross the ridge through a wide, open saddle cut 500 feet below the ridge crest. This is a wind gap, the abandoned valley of a stream that formerly flowed across the ridge. The pumping installation, originally named Wind Gap plant, is now the I. J. Chrisman plant, honoring a California Water Project Engineer.

South of, but separate from, Wheeler Ridge are the smooth, grassy slopes of Pleito Hills. Rising to a high and ragged skyline behind is the rugged San Emigdio Range, here culminating in Mt. Tecuya at 7,150 feet. Wheeler Ridge is a separate entity, rising more than 1,000 feet above the smooth apron that slopes gently northward from the higher terrain to the south. What is it doing out there all by itself, a sort of isolated maverick?

Low-level, oblique air photo (taken in 1957, before construction of the aqueduct) of east end of Wheeler Ridge viewed from the south. The wind gap lies a little west of center. The floor of San Joaquin Valley fills the middle ground, with the Tehachapi Mountains in the distance. Salt Creek water gap and related dissection are visible in the lower right. —John S. Shelton photo

Wheeler Ridge may be isolated, but it is no maverick. It is one of the best examples anywhere of an upward-arching fold, called an anticline, so young that the shape of the ridge replicates the arching form of the fold—modified only slightly by gullies on its slopes, one water gap, and several wind gaps indenting its crest. A water gap is a canyon cut through a ridge or mountain, which is still occupied by a stream.

Wheeler Ridge stretches 8½ miles from east to west and is 3½ miles wide at most. Like the curved back of a sounding whale, it dives beneath the ground at its east end and tails off into Pleito Hills to the west. Oil and gas, if they exist in the structures, collect in anticlines, so oil companies have drilled many holes into Wheeler Ridge. These holes and those in the adjacent Tejon and North Tejon oil fields reveal that the rocks underground are considerably more complex than the simple surface anticline would suggest. A large thrust fault that dips south cuts the fold along its length, and at greater depths, other thrust faults in older rocks further complicate the picture. A thrust fault is a gently inclined fracture surface along which the upper block of the earth's crust has moved up relative to the lower block.

If you followed directions and are now in the wind gap, you can see boulders littering the slopes of Wheeler Ridge. They weathered out of the Tulare formation, a deposit of still unconsolidated coarse sand and gravel laid down by streams as recently as 250,000 years ago. The Tulare formation caps the sequence of beds folded into the Wheeler Ridge anticline. Although bedding within the formation is crude and hard to see, you can probably tell that it tilts gently down to the south. The anticlinal crest lies north, so backtrack by car or foot about 0.3 mile down the road to a large west-side gully with bare slopes and cliffs at its head. There you can see the change in direction of bedding inclination that defines the crest of the anticline.

Turn your car around at a graded flat near the curve, 150 yards down the road. Return to the aqueduct, cross it, and continue south.

The boulders from the Tulare formation are worth a look. A good collection lies at the base of the slope across the canal, just beyond the cattle guard. They are hard igneous and metamorphic rocks, with an occasional chunk of marble, derived from distinctive rocks in the San Emigdio Mountains on the southern skyline. This part of the Tulare formation consists of coarse debris that floods swept out of those mountains before Wheeler Ridge existed. Most of the boulders are angular or subangular, their sharp edges blunted so slightly that they could not have traveled far. The maximum size of the boulders in this vicinity is six feet. These deposits differ little from sand and gravel in the beds of streams that now drain from the San Emigdio Mountains.

Continue south on the wind gap road, keep southwest (right) on the wide, well-graded branch at the first intersection, 0.3 mile beyond the aqueduct crossing, and then turn east left at the T intersection in another 0.3 mile. Proceed 1.4 miles and turn north (left) on a crudely macadamized road leading into the water gap.

This road is a bit rough, but not hazardous, and the drive back across Wheeler Ridge is instructive. The Tulare formation is well exposed in the canyon walls, its bedding obvious enough that you can see it wrapping across the arching crest of the anticline. The fold is not as sharp here as it is farther west, and its north limb is not as steep. The road through the water gap emerges beside a large sand and gravel operation, which mines the Tulare formation and associated deposits. The hard, fresh boulders from the Tulare formation make excellent commercial gravel when crushed.

Northbound travelers can continue north to the intersection with a heavily traveled road bearing northwest toward the Chrisman pumping plant. Follow it to Sabodan Street, which will lead you north to either freeway. Travelers wishing to go south on Interstate 5 can turn southeast (right) at the gravel mine, drive to the frontage road paralleling the freeway, and continue south to an on-ramp onto Interstate 5 in about 1.25 miles.

We have been too close to Wheeler Ridge to see much of the features along its crest. Air photos and topographic maps show several shallow wind gaps indenting the crest west of the aqueduct. These are clearly visible to travelers on California 166 within the first 1.5 miles west of its overhead crossing of Interstate 5. The situation here was similar to that in the Puente Hills (Vignette 9), where San Gabriel River created Whittier Narrows. Initially, a number of streams flowed north out of the San Emigdio Mountains and Pleito Hills, across a gently sloping alluvial apron to the floor of the San Joaquin Valley. Wheeler Ridge anticline started to arch across the paths of these streams. Smaller streams and those facing the most rapidly rising part of the ridge were shunted, after they cut only shallow valleys across the rising fold. These abandoned valleys became wind gaps as the anticline continued to rise.

The stream that eroded the crossing for the aqueduct was near the younger and less rapidly rising eastern end of the fold, where it was able to cut down as rapidly as the barrier rose across its path. Water from some of the diverted streams probably augmented its flow. The "aqueduct stream" maintained its course across the anticline during about the first half of its rise. Then it failed to keep pace. Perhaps the ridge began to rise faster than the stream could erode its bed, or possibly the increasing width of the fold defeated the stream. Whatever the reason, the rising fold shunted the stream to the east, where it still discharges through the water gap. As the ridge continued to rise, the deep gash our "aqueduct stream" eroded became the largest and most spectacular wind gap on Wheeler Ridge.

Many years later, aqueduct engineers, seeking an inexpensive means of crossing Wheeler Ridge at a convenient elevation, spotted the wind gap. It was at the right height for water to flow by gravity to the huge Edmonston pumping plant where it is boosted nearly 2,000 feet to cross the big faults of the Tehachapi Mountains on the surface. Using the wind gap capitalized on the stream's erosional work as it battled the rising anticline. Aqueduct engineers should periodically honor that valiant stream, gone but not quite forgotten.

Erosion of Wheeler Ridge, mostly by gullies on its flanks, is considerably less than that in the Puente Hills (Vignette 9) and much less than at the Ventura anticline (Vignette 4), suggesting that it is a younger structure. Young stream deposits at its east end are tilted, which suggests that the fold may still be rising, but no tiltmeter or leveling surveys have continued long enough to provide direct proof. The tilted stream sediments suggest that the fold is rising at a rate of about one-sixteenth of an inch per year. At that rate, it would have taken about 200,000 years to raise the entire structure. Engineers need to worry about the current uplift; a small change in elevation can play havoc with the aqueduct gradient.

The Arvin-Tehachapi earthquake of 1952, magnitude 7.5, came from a focus, the point within the earth's crust where the shock originated, directly beneath Wheeler Ridge. That provides another reason to think the anticline may still be rising. Surveys made half a year after the 1952 earthquake found that Wheeler Ridge had been lifted as much as two feet since the last survey, made a year earlier. This was before construction of the aqueduct.

Wheeler Ridge is a beautiful example of a young anticline, probably still growing, clearly visible to motorists on two of southern California's busy freeways. Its wind gap is spectacular. The grass there is green in a wet spring, and the smooth slopes south of Wheeler Ridge provide fine conditions for wildflowers.

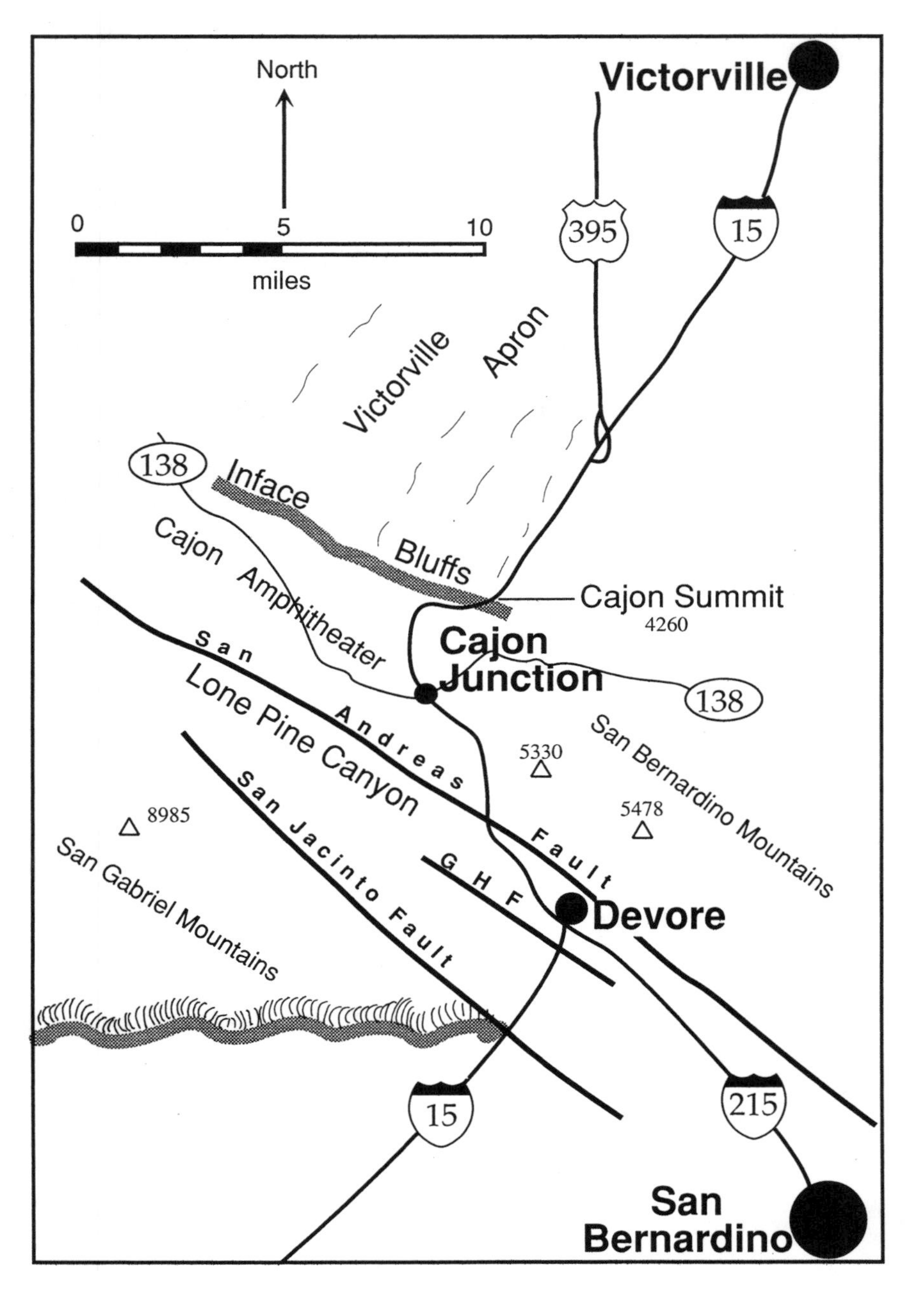

Highways and features of Cajon Pass. GHF is Glen Helen fault.

VIGNETTE 11

THE SAN ANDREAS FAULT AND CAJON CREEK COOPERATE

Cajon Pass

SAN BERNARDINO COUNTY

Every day thousands of vehicles and people cross Cajon Summit on Interstate 15. Three rail lines, four power lines, and several pipelines also use the pass. Cajon Summit, the high point on the highway, is about two miles northwest of the true pass, which the railroads cross. Motorists have a good chance of seeing a many-engined train climbing or descending the pass. In earlier days, Indians, explorers, the Spaniards, and the Mormons used this path between the desert and the coastal regions. Perhaps a few of the thousands who cross Cajon Summit wonder about the combination of circumstances that created this gateway.

Were it not for Cajon Pass, Interstate 15 and the railroads would need to climb at an 11 percent grade to at least 5,000 feet, within a scant five miles from the south base of the mountains. As it is, the railroads clear Cajon Pass at 3,880 feet, 16 miles from the mountain base to the south, after climbing a 2 percent grade. The highway crosses Cajon summit at about 4,260 feet, 13 miles from the south base, with a grade averaging 3.25 percent. Cajon Pass is a gift from nature that merits our appreciation.

After climbing a long grade, Interstate 15 approaches Cajon Summit from the south by ascending obliquely across the face of some steep slopes that Levi F. Noble (1882-1965), a prominent early geologist, called the "Inface Bluffs." Beyond the summit, the highway starts its long, gradual northward descent toward Victorville, seventeen miles ahead. On the skyline to the south and southwest, the rugged crests of the San Bernardino and San Gabriel mountains, attractively snowcapped in winter and spring, rise to 10,000 feet. Between Cajon Summit and this high country is a broad, open valley of much lower elevation and gentler relief, the Cajon amphitheater. The Inface Bluffs bound the amphitheater's north edge. Sloping gently northeastward from Cajon summit is a smooth surface of sand and gravel, the Victorville apron, the southern margin of which ends abruptly at the Inface Bluffs.

Numerous cobbles of a hard, dark gray, slabby rock, a nodular mica schist, exist within the gravel beneath the Victorville apron and on its surface. They are typical of the Pelona schist, a common building stone in southern California. As the cobbles weather, they become brownish, and the mica, still shiny, turns golden. Pelona schist is widely exposed in the San Gabriel Mountains to the southwest. These fragments presumably came from there. But how did they cross the low ground of the Cajon amphitheater? There is no existing downhill route the cobbles could follow from the San Gabriel Mountains to the Victorville apron. That is anomaly number one.

Oblique air view west-northwestward into the Cajon amphitheater. The dissected Victorville apron slopes right (north); the Inface bluffs face left (south) through mid photo. The high San Gabriel Mountains occupy the photo's upper left. Railroads cut across the lower left, and Interstate 15 curves through the middle. —John S. Shelton photo (taken in 1961)

Beheaded, hanging valley truncated by Inface bluffs west of Interstate 15.
—John S. Shelton photo (taken in 1960)

A topographic map or air photo shows that the upper reaches of the Victorville apron are indented by broad washes, called arroyos, incised some 100 feet into its surface. Streams that build smooth alluvial aprons with a slope nicely adjusted to carry their water and sediments do not ordinarily turn around and dissect that apron without cause. They would do so only after some independent perturbation such as a change in climate or a rise or tilt of the land.

Extremely precise geodetic surveys of the Cajon Pass area show that this region has risen with respect to the country to the north at a rate of about one-half foot per century, during the period of the surveys. It is reasonable to suppose that the land has been rising for a longer time and that its rise caused the washes to dissect the Victorville apron.

The washes are peculiar in other ways, however. If you were to walk up one of them—Manzanita Wash about four miles west of Interstate 15—you would find it extends with full width and depth right to the brink of the Inface Bluffs. There it stops. Its upper part is gone. It is a beheaded wash, as are the others. That is anomaly number two. The best clues to an explanation of these anomalies and to the origin of Cajon Pass involve the history of Cajon Creek.

Cajon Creek separates the San Gabriel and San Bernardino mountains. They are part of the Transverse Ranges, so called because of their westerly trend, transverse to the general northerly trend of the mountains in southern California. These west-bearing mountains form an essentially continuous wall along the inland side of the coastal lowland, which Cajon Pass breaches.

Cajon Creek starts its penetration of the mountains at Devore, where Interstate 215 joins Interstate 15. Most streams flowing out of the mountains follow courses roughly perpendicular to the mountain front. Cajon Creek and its next door neighbor to the west, Lytle Creek, do not. They flow southeasterly, obliquely to the mountain front. Both follow the trends of swarms of fractures that parallel two of the largest faults in southern California, the San Andreas and the San Jacinto. In its lower part, Cajon Creek lies midway between these faults and follows the subsidiary Glen Helen fault. Streams are masters of finding easy rock to erode, and they rarely miss the crushed and ground rock in a major fault zone.

As Cajon Creek canyon penetrates the mountains, it becomes narrower and deeper, and Interstate 15 rises higher onto its northeast wall. The San Andreas fault lies a scant mile or two to the northeast, on a bearing that gradually converges with the highway. About three miles beyond Devore, the highway curves broadly to the northeast, and enters the first of two deep, double-walled roadcuts. Much of the rock in these and some of the preceding roadside cuts is laminated, dark gray Pelona schist. It is a major part of the rock terrane southwest of the San Andreas fault. The rock in the last cut is dark and slightly greenish, and much broken by the stress and strain of being alongside the San Andreas fault, possibly for millions of years.

The change in highway bearing puts the road on a collision course with the San Andreas fault. They meet just north of the double-walled cuts. A quick look to the west, under favorable lighting and visibility, gives a fine view up narrow, abnormally straight Lone Pine Canyon, aligned along the trace of the San Andreas fault zone. Lone Pine Canyon is anomalous in that it cuts across drainages flowing out of high country to the southwest. The canyon looks like a discordant scar hacked into the face of the earth by a giant machete.

Like the interstate, Cajon Creek curves northeast to meet the San Andreas fault. That is where Cajon Canyon changes from a narrow gorge to an open valley. The gray Pelona schist and related rocks southwest of the fault resist erosion, so the creek passes through them in a steep-walled valley. The same creek eroded a much wider valley in the softer and more colorful sedimentary rocks northeast of the fault, mainly the Cajon Valley beds. Watch the valley slopes to see shades of cream, pink, and tan in the Cajon Valley beds.

At Cajon Junction, where Interstate 15 meets California 138, the highway enters the wide, open Cajon amphitheater. To the west, watch for the prominent ridges of resistant Cajon sandstone, a pale rock. These are locally known as Mormon Rocks, memorializing early Mormon pioneers who habitually camped along Cajon Creek on their way from San Bernardino, which they founded, to Utah.

Cajon amphitheater is the key to easy passage through the mountains. Why does it exist? What sequence of events and processes displaced the drainage divide between coastal and desert areas five miles northeast and lowered it at least 1,000 feet? The San Andreas fault and Cajon Creek played major roles.

In geological parlance, the San Andreas is a right-lateral fault. The blocks on opposite sides slip past each other horizontally. No matter which side you are on, the opposite side appears to move to the right.

The shifted and lowered divide at Cajon Pass.

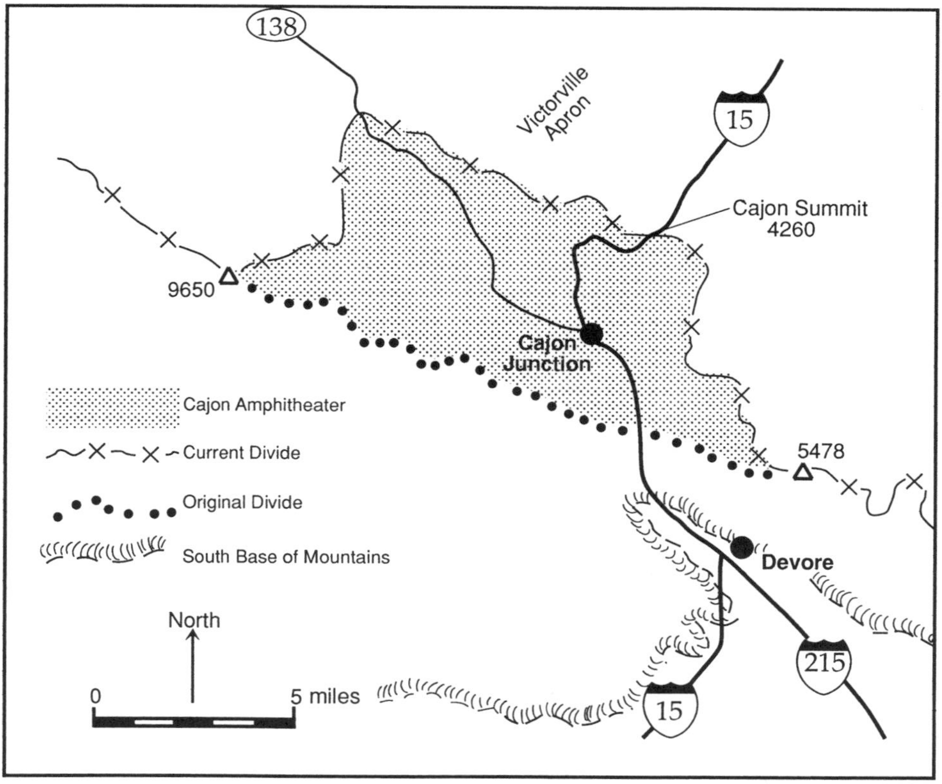

Many geologists accept evidence suggesting that its total displacement in southern California is at least 190 miles, with the rocks west of the fault moving north. Some geologists think the total movement may amount to as much as 340 miles. Movement on the San Andreas fault caused the very large 1857 Fort Tejon earthquake and the more disastrous 1906 San Francisco earthquake. The fault separates the North American and the Pacific plates of the earth's crust, no trivial function.

The San Andreas long ago moved relatively soft and unconsolidated sedimentary rocks, the Cajon Valley beds, into the area that is now the Cajon amphitheater. The Cajon Valley beds are six to 15 million years old, and they moved into this area before any of the present topography existed.

Cross section and map A show the topographic setting as it existed roughly half a million years ago. The south face of the mountains rose steeply from the floor of San Bernardino Valley to a ragged crest that was probably more, perhaps considerably more, than 5,000 feet above sea level. The divide between drainages to the coast and desert lay along this crest. A gently sloping alluvial surface, the Victorville apron, spread northeast from high on the desert face of the mountains across the area now occupied by the Cajon amphitheater. Its veneer of sand and gravel covered the Cajon Valley beds. Since this alluvium was derived from the mountains, it contained fragments of their rocks, particularly chunks of Pelona schist.

Anyone attempting to cross the mountains from the coastal side before Cajon Pass formed would have had to climb about 4,000 up the steep south face of the mountains to an elevation above 5,000 feet before dropping to the head of the alluvial apron for the smooth passage northeast to Victorville. Someone traveling in the opposite direction would have had to climb at least 1,000 feet higher than at present and then would have faced a very steep descent to the San Bernardino Valley. It would have been like crossing the San Bernardino Mountains today near Crestline, southeast of Cajon Pass.

While streams flowing from the northeast side of the mountains were building the Victorville apron, Cajon Creek was eroding its canyon. By the time Cajon Creek extended five miles into the mountains, the San Andreas fault lay only about a mile to the northeast. Like any prudent explorer, Cajon Creek sent out advance scouts—tributaries. One tributary extending to the northeast discovered the wide zone of macerated and very erodible rock within the San Andreas fault zone. The tributary soon cut across it to the soft Cajon Valley beds. This tributary rapidly eroded its valley and soon began to capture the headwaters of streams flowing northeast from the mountains into the desert. This so augmented its water supply that it quickly became the main trunk of Cajon Creek.

Once Cajon Creek entered the soft rocks, its tributaries were like

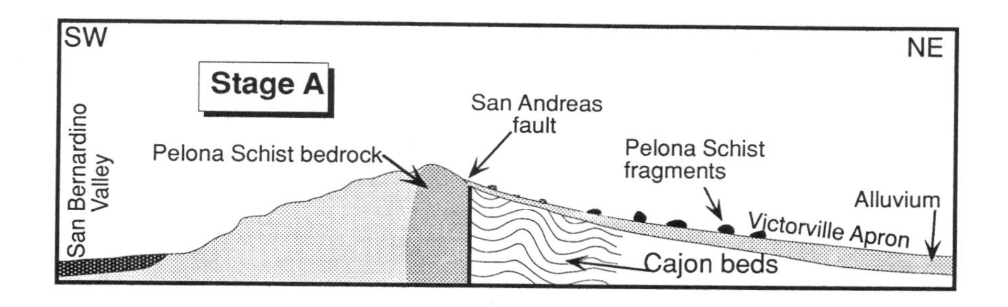

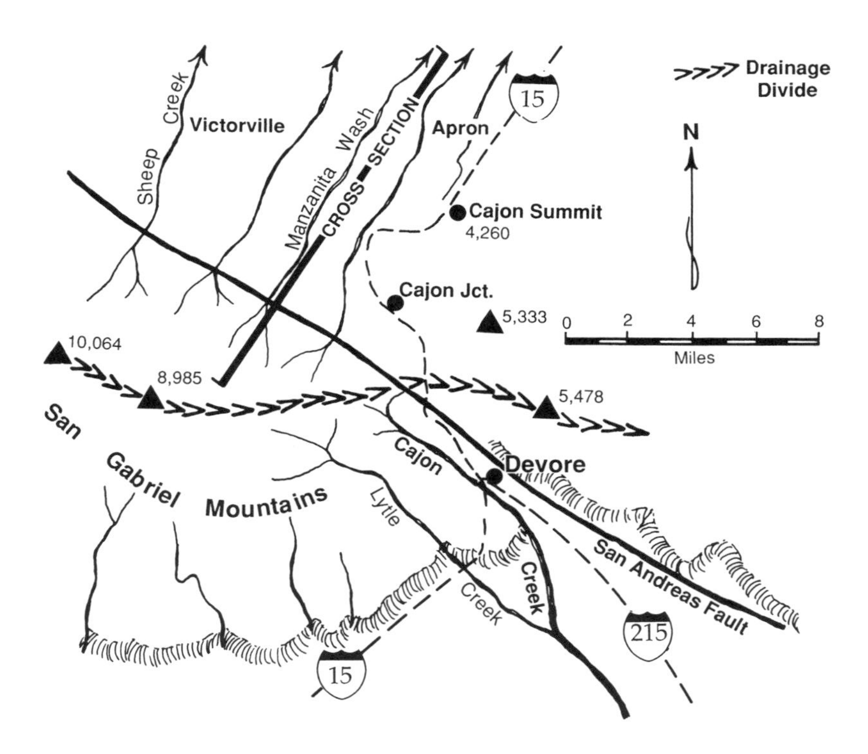

Stage A. San Andreas fault has moved soft Cajon beds into position (see cross section), but they are buried by a veneer of alluvial gravel (see cross section) containing Pelona schist stones. Cajon Creek has penetrated the mountains and sent a tributary northeast almost to the San Andreas fault zone. The drainage divide runs through high peaks of the mountains, and the Victorville apron extends to their foot.

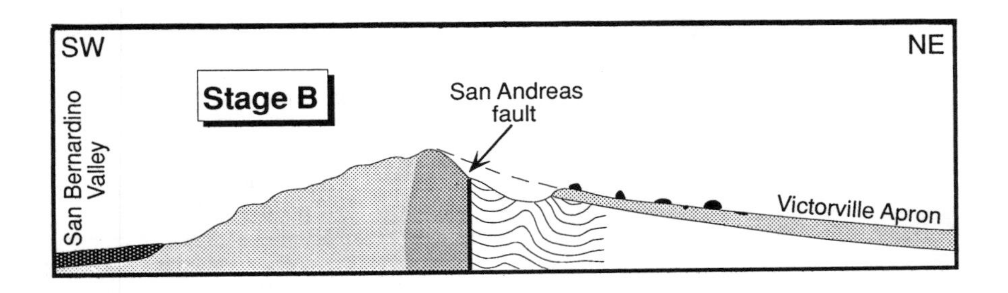

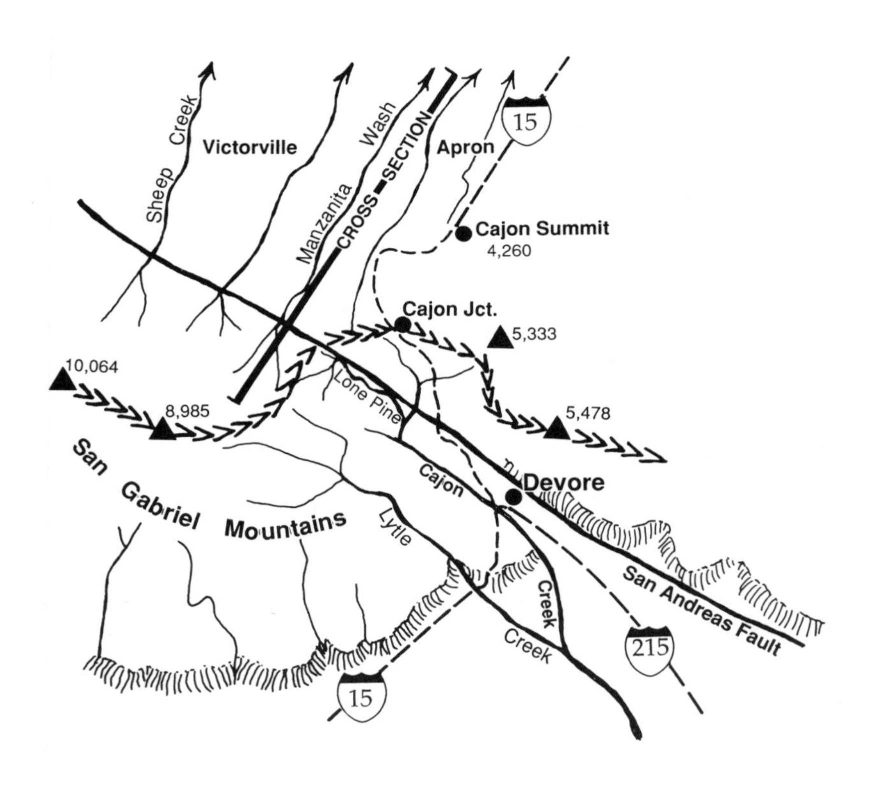

Stage B. Cajon Creek crosses the San Andreas fault and erodes soft Cajon beds, driving the drainage divide north. Lone Pine, a tributary to Cajon Creek, works headward along the San Andreas fault zone. At least one Victorville apron stream is beheaded.

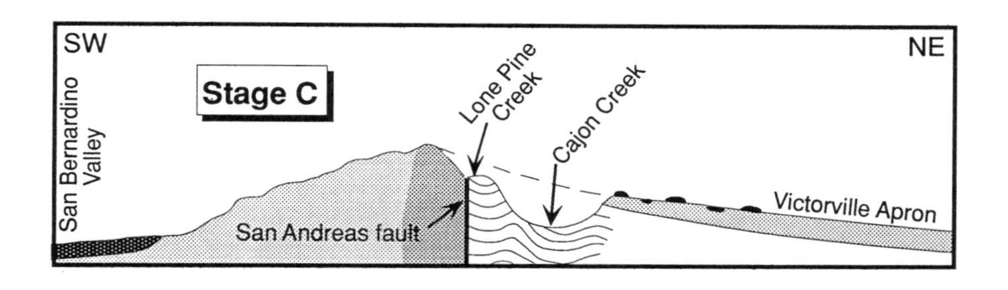

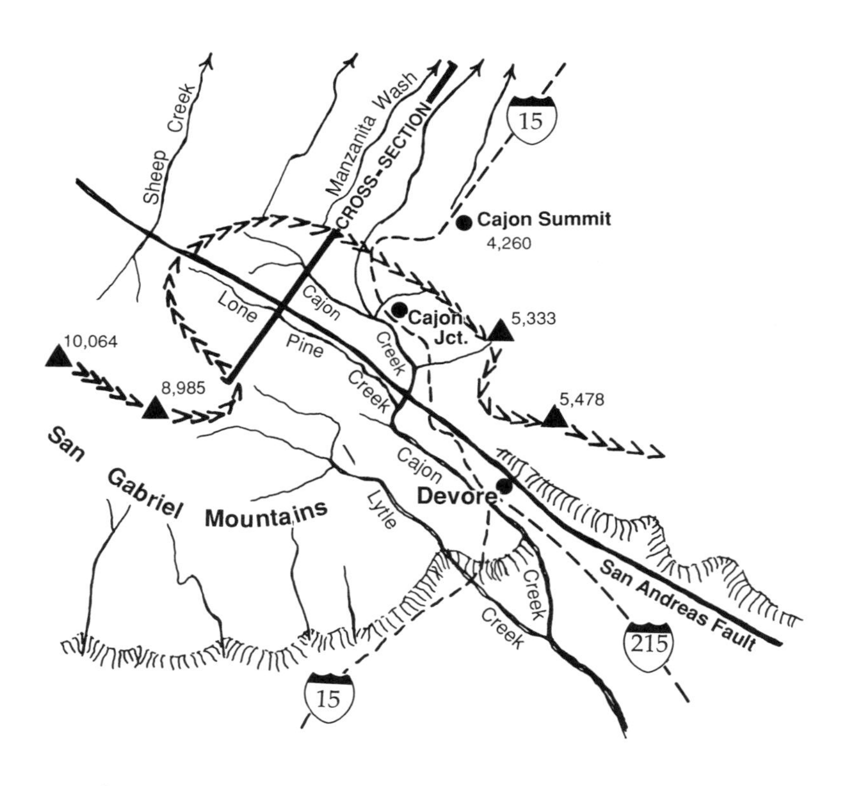

Stage C. Cajon Creek enlarges the erosional amphitheater it is creating in the Cajon beds. With the help of its parallel tributary, Lone Pine Creek, Cajon Creek drives the drainage divide far to the northwest. Incipient Inface bluffs are probably forming.

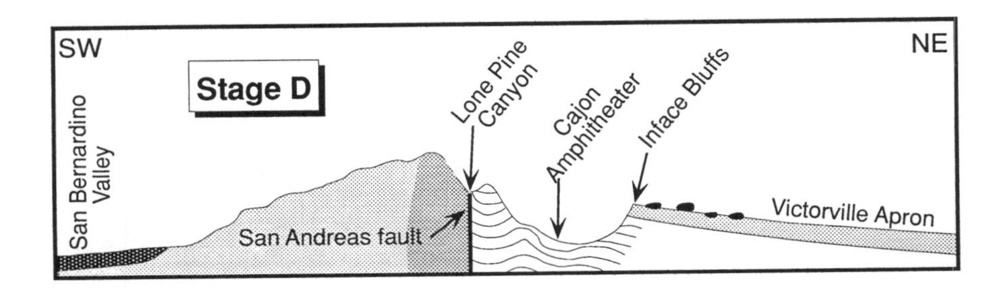

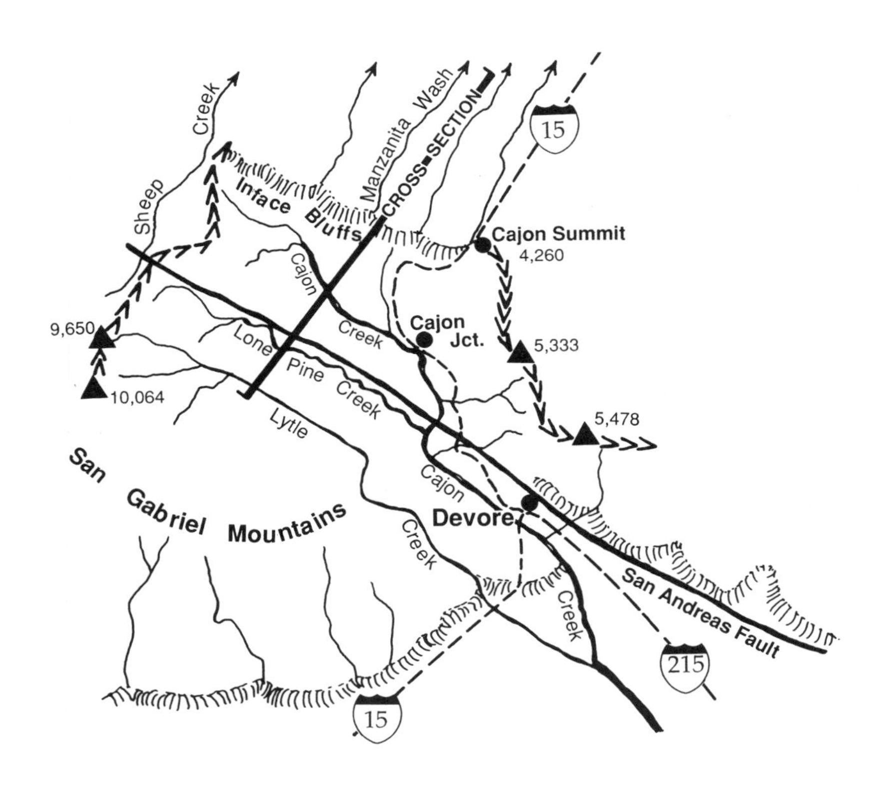

Stage D. Continued erosion by Cajon Creek and tributaries has created the Cajon amphitheater and the Inface bluffs as they exist today. The drainage divide has been driven 5 miles northeast, and the crest of Cajon Pass has been lowered from over 5,000 feet to 4,260 feet above sea level.

voraciously eating into the soft sedimentary beds and creating the Inface Bluffs. Gullies eroding the bluffs' steep face and streams and ground-water sapping at their base ate into the Victorville apron and caused the bluffs to retreat northeastward. This drove the drainage divide to the northeast and also lowered it, because the apron slopes down in that direction. The arroyos on the apron were cut off from their former supply of water and debris in the San Gabriel Mountains. These gullies were left hanging into empty air at the Inface Bluffs, literally beheaded. Recession of the drainage divide, defined by the brink of the Inface Bluffs, between streams flowing northeast and southwest continues slowly today. To visualize this sequence of events, study carefully the cross sections, maps, and captions in the accompanying illustrations.

An explanation for the two anomalies discussed earlier now becomes clear. The washes on the Victorville apron once extended several miles farther southwest to the base of the San Gabriel Mountains. They have been beheaded by the formation and recession of the Inface Bluffs under the erosional attack of Cajon Creek tributaries, thus producing anomaly number two. As for anomaly number one, the Pelona schist fragments in Victorville apron alluvium were carried from the mountains and spread across the Victorville apron before Cajon amphitheater existed. These anomalies may seem to be relatively minor items, but they provide important clues to the succession of events that created Cajon amphitheater and Cajon Pass.

Although Cajon Creek is the agent most directly responsible for carving the amphitheater, it could never have done the job if soft sedimentary rocks had not already been moved into place by San Andreas fault. The San Andreas fault created the critical relationships that Cajon Creek exploited.

So, the next time you escape crowded southern California by way of Cajon Pass for a vacation in the Sierra, the desert, or a weekend in Las Vegas, give the San Andreas fault a friendly nod in passing. It is your benefactor as well as your enemy.

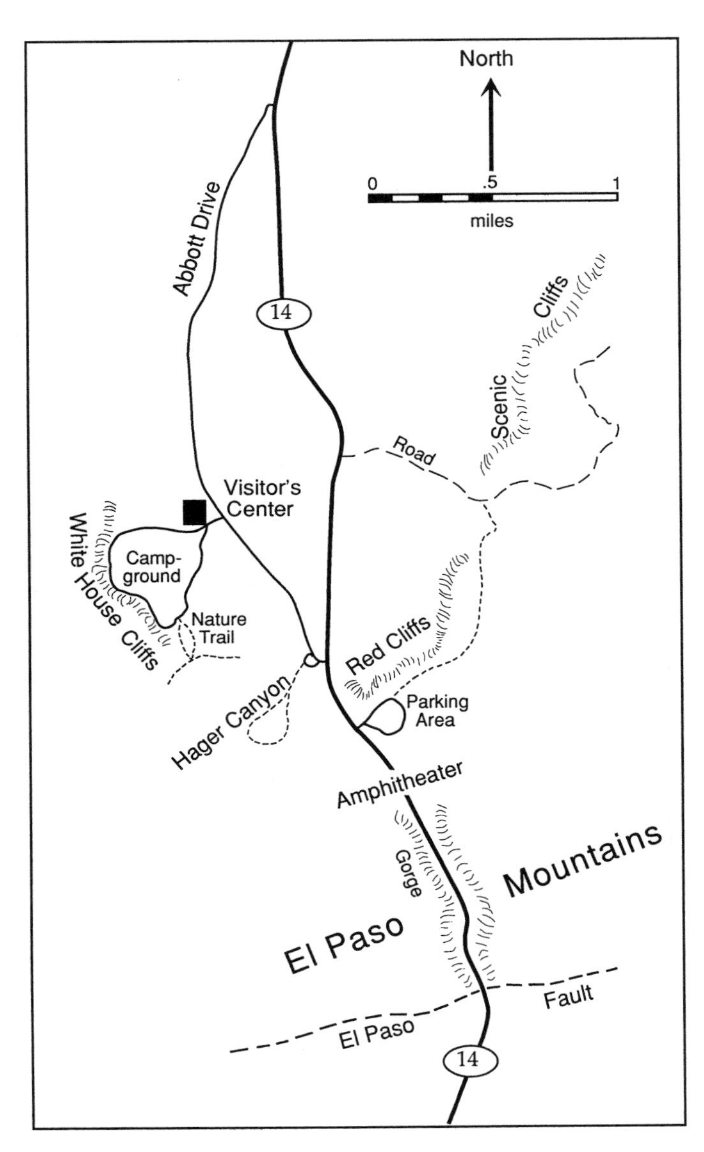

Red Rock Canyon and vicinity.

Auto club maps of Kern County show highway accesses to Red Rock Canyon. Geographic details appear on U.S. Geological Survey, 7.5-minute, topographic quadrangles Cantil, Saltdale NW, Dove Spring, and Cinco, of which the first two are the most useful.

VIGNETTE 12

RED ROCK CANYON

A Geologic Library

KERN COUNTY

Millions of people from all over the world have seen the brightly colored, badland cliffs of Red Rock Canyon in the Mojave Desert, twenty miles northeast of Mojave on California 14. The castellated cliffs provided background for early cowboy films of the William S. Hart and Tom Mix era and for many other Hollywood movies. You saw them in the TV show, "Bonanza," and they appear regularly in TV commercials. Thousands of motorists see them every day as they drive north from heavily urbanized southern California to the vacation areas of the eastern Sierra Nevada. California 14 passes through the heart of the canyon.

Red Rock Canyon is famous scenery. Geologists enjoy the scenery too, but they treasure the canyon and its rocks even more as an archive of information about the history and evolution of the surrounding country. To them, it is a library, its sedimentary beds being like pages in books. Reading those pages requires skill, knowledge, and experience. Let us try to show you how it is done.

Badlands are bad only in the sense that their steep and barren slopes make them useless for grazing and difficult to cross. We all tend to detour around them. But the scenery of badlands is invariably striking and commonly colorful. Bryce Canyon and Cedar Breaks in Utah and the Badlands of South Dakota are good examples. Smaller badland areas dot the arid southwest, and many are protected as county or state preserves. Red Rock Canyon is part of a state park established in 1970.

Badlands most commonly form in sedimentary deposits soft enough to erode easily, but coherent enough to stand in very steep faces. Rain splash is the dominant process of erosion. It is a highly effective and powerful process, especially where heavy rains fall on barren slopes that shed surface runoff in threads, rills, and gullies. Rill and gully channels can be so deep, steep sided, and narrow that stout people cannot easily pass through them. Nearly every badlands has its "Fat Man's Misery."

Rain splash and surface runoff carve a geometrical collage of columns, spires, flutes, chutes, knife-edge ridges, niches, alcoves, and sundry other forms. Variations in color, including white, cream, beige, brown, pink, red, and green, within different layers add to the scene. A varied coloration usually reflects a high content of volcanic debris in the eroding sedimentary formation, much of it rhyolite ash. This is especially true in Red Rock Canyon.

Highway 14 makes a dramatic entry to Red Rock Canyon from the south. After passing northeastward across some twenty miles of the broad and gently sloping alluvial apron that flanks the south end of the Sierra Nevada, the highway curves northwest. It then heads directly toward the abrupt south face of El Paso Mountains, the scarp of El Paso fault. Watch for the fault which is nicely exposed at the mountain base a few hundred yards east of the highway. The road soon enters a narrow

Castellated "accordion folds" cliffs at west end of parking area, east of California 14. Massive pink volcanic tuff-breccia forms left skyline. Foreground pebbles weathered out of conglomeratic beds in cliffs.

Red Cliffs at parking area in Red Rock Canyon amphitheater, just east of California 14. Massive pink volcanic tuff-breccia caps the cliffs.

gorge in dark brown, extremely fractured granite. After passing through nearly a mile of granite, the road bursts into the Red Rock amphitheater, and striking badland cliffs come into view.

For a leisurely inspection, turn right off California 14 into a large parking area, with restrooms, 1.25 miles from the gorge entrance. There, to the north, are the Red Cliffs, noted for their closely spaced drainages that make them look almost like accordion pleats, especially toward the west end. Near the base of the cliffs, the ground is littered with small pebbles weathered out of the sedimentary rocks. Most are fragments of a wide variety of volcanic rocks, but a few are granite or metamorphic rocks. These stones contribute to the spectrum of colors in the cliffs, but the large content of volcanic ash contributes more.

At first glance, the sedimentary layering appears horizontal, but a more careful look shows that it tilts away at an angle of about 15 to 20 degrees. The eroded forms follow sedimentary layers to some extent, but most cross them.

Walk south across the parking area until you can see the top of the bluffs. A thick and massive layer of tan to pink rock caps them. It is much fractured and weathers to subdued knobby surfaces. Highway 14 passes

Close-up view of volcanic tuff-breccia, with pocket knife (3½ inches long) for scale.

Fault, to left, inclined right, displaces contact of massive pink tuff-breccia and underlying sedimentary layers, just east of California 14 and a little north of Red Cliffs.

through this layer shortly north of the parking area. You can lay hands on the bluff-capping layer by circling the west end of Red Cliffs and walking 200 yards north. Close up, you see that a fine matrix of consolidated volcanic ash, or tuff, encloses small angular fragments of a variety of rocks, mostly volcanic, which explains why this type of rock is called tuff breccia. This tuff was deposited as a fast-moving, hot ash flow, erupted from a nearby volcano 12 to 13 million years ago. That volcano was probably in the Mojave desert on the south side of the Garlock fault, which has since moved it many miles eastward. Right in front of you is a small fault, inclined steeply east, that displaces the contact between the tuff breccia and the underlying sedimentary beds by about thirty feet.

To see more of Red Rock Canyon, leave the parking area, head northbound on California 14, and turn west (left) onto Abbott Drive. Follow it 2.5 miles north to intersect again with California 14, labelled the Midland Trail. This is the old Red Rock highway, and it takes you past the entrance to the park ranger station, Visitor Center, and campground. The Visitor Center offers exhibits, as well as maps, publications, and information on many aspects of human and natural history.

A short nature trail starts from the campground road near the last group of campsites at the base of White House Cliffs, just beyond campsite 46. This is an easy loop, on a well-marked path, with a dozen numbered cairns and a trail guide. You can extend the walk at the short flight of steps near cairn 12. Follow the broad trail south, up the ridge to a bench, and look out into the Mojave Desert and Fremont Valley. Trail branches lead south into Hager Canyon or east up a steep ascent to the top of a lava-capped ridge for a great view of Red Rock Canyon and larger parts of the Mojave Desert.

Another easy and well-marked trail follows a mile loop in Hager Canyon. It starts from a spacious parking spot at the south end of Abbott Drive. From it, you can wander into several side canyons and see other exposures of the massive pink bed of tuff breccia.

A slightly more demanding walk north out of the Red Cliffs parking area follows a broad trail, actually an abandoned road, that parallels the Red Cliffs. In a little less than a mile, it meets a traveled road and enters a broad valley that ends in a splendid amphitheater a mile farther north. This is the Scenic Cliffs area. A network of old roadways, foot trails, and flat-floored washes make wandering and exploration easy. Besides its engaging badland scenery, this area displays faults cutting beds in the cliffs, massive landslides, especially involving the pink tuff breccia in both the Red and Scenic cliffs, and some dark basalt dikes that cut through the Ricardo group of sedimentary rocks. The traveled road ends near the mouth of Nightmare Gulch, a narrow defile with steep walls.

About a half mile along the trail from the Red Cliffs parking area, the top of a long, narrow ridge to the east has pale beds of the Ricardo group

Cliffs in the Dove Spring formation of the Ricardo group, capped by interbedded lava flows, just west of California 14 in Red Rock Canyon amphitheater.

lying on tilted layers of reddish rock. Although it looks like one, this is not a normal angular unconformity. An angular unconformity is a break in the geologic record in which younger sediments rest upon the eroded surface of tilted or folded older rocks. The pale layers of the Ricardo group above the tilted beds are possibly a remnant of an old landslide that was isolated on the ridge crest as streams eroded canyons on both sides.

The traveled road into the Scenic Cliffs area branches from California 14 a bit more than a mile north of the road into the Red Cliffs parking area. No sign identifies this junction, but it is easy to spot where a straight reach of California 14 bends into a broad curve to the west. The condition of the road into the Scenic Cliffs area makes it easier and more fun to walk from the Red Cliffs parking area. The round-trip of four miles is well worth a half day, even a full day. This is one of the best parts of Red Rock Canyon, scenically and geologically, and remote enough to lend a sense of adventure.

Red Rock Canyon has fascinated geologists ever since one of the greatest of the breed, G. K. Gilbert (1843-1918), gave it a complimentary nod in an 1875 report on the geology of the western United States. Discovery of fossil mammal bones increased interest during the early 1900s. Scientific papers on the area continue to appear regularly.

Most large regions can be conveniently subdivided into geologic provinces. Southern California has nine. Three—the Mojave block, the Sierra Nevada, and the Basin and Range—come to a triple junction at Red Rock Canyon. The canyon had a ringside seat at happenings within those provinces, especially during the last 19 million years. That history is recorded within the sedimentary layers of the Ricardo group of formations, a name derived from the only enduring settlement that ever existed within the Canyon, the Hager family establishment. Ricardo was either an innkeeper there, a small boy who tended the horses of the stagecoach that traveled between Mojave and Bishop, or a member of the Hager family; authorities disagree.

The Mojave block is shaped like a wedge, with a sharp western point at Lebec and broadening east to the Colorado River. The San Andreas fault bounds its southwest side; the Garlock fault zone marks the north side. El Paso fault is a part of the Garlock fault zone. The Sierra block trends north and south, west of the Sierra frontal fault system, which passes along the west side of the Red Rock Canyon area and terminates against the Garlock fault. Red Rock Canyon actually lies within the southern part of the Basin and Range province, which contains a succession of lofty mountain ranges that trend north, separating deep basins such as Owens, Panamint, and Death valleys. In the 1880s, geologist Clarence E. Dutton described the parallel mountain ranges of the Basin and Range as "looking upon the map like an army of caterpillars crawling northward out of Mexico." Rocks exposed in Red Rock Canyon record events that occurred within these three provinces, especially along the Garlock and Sierran fault systems.

The Cudahy Camp formation and the overlying Dove Spring formation make up the Ricardo group of formations, with a total thickness of some 6,500 to 7,000 feet of sedimentary layers. In Red Rock Canyon, we see only rocks of the Dove Spring formation. The Cudahy Camp formation is exposed principally in Last Chance Canyon, farther east.

During Early to Middle Tertiary time, 55 to 19 million years ago, the Mojave block was eroding, shedding great quantities of sediments south and west into adjacent marine sedimentary basins. Near the end of this period, the geological gremlins in charge of the Mojave block apparently decided it was unwise to export all that debris; some should be kept at home. Perhaps they recognized that erosion is the great eraser, deposition the great preserver, of geological records. Small, local basins, where records of events could be preserved, formed by warping and faulting within the Mojave block. The Barstow basin (Vignette 13) is one of these; our El Paso basin, another.

Sedimentary layers that accumulated in El Paso basin from 19 to something less than 7 million years ago became the Ricardo group of formations. Their cumulative thickness of around 7,000 feet does not mean that a basin of that depth formed first and then filled to the brim. Rather, the basin sank slowly, bit by bit, and sedimentation kept pace, so

Tectonic setting of Red Rock Canyon, showing its strategic location at triple junction of the Sierra Nevada, Basin and Range, and Mojave geologic provinces.

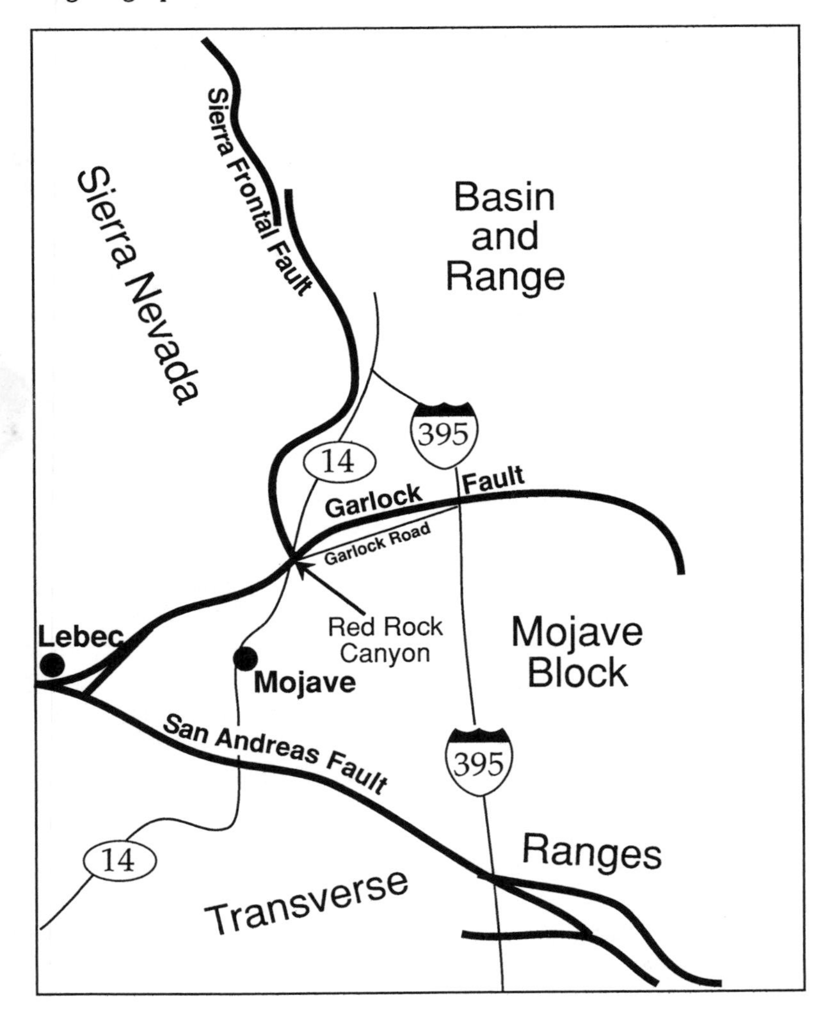

the basin was always nearly full. Long before the Ricardo episode of filling, 60 to 55 million years ago, El Paso basin had accumulated an even greater thickness of sediments, the 14,000-feet-thick Goler formation. Those layers were tilted, then eroded before the Ricardo formations were deposited. The angular unconformity thus created marks a 35-million-year hiatus between deposition of the two formations. The Dove Spring formation is simply the top of a very thick sedimentary column filling El Paso basin.

A shallow, possibly ephemeral, lake occupied part of El Paso basin during the late deposition of the Dove Spring formation. The upper Dove Spring beds contain considerable fine lake and floodplain sediment, within which are the fossil remains of a rich fauna of mammals. The fossils include early horses, camels, mastodons, rhinoceroses, wild dogs, pronghorn antelopes, deer, sabertooth tigers, cats, weasels, one species of goose, and possibly a skunk or two. The grazing must have been good. Although grasslands probably predominated, trees also grew. The formations contains fossils of pinyon pine, locust, cypress, acacia, and palm trees. Petrified palm wood is especially attractive when polished. The annual rainfall may have been around 15 inches, three times the canyon's present quota but similar to that of present-day Los Angeles. All these fossils suggest that the upper Dove Spring formation is about ten million years old.

In 1988, two young geologists, Dana Loomis and Douglas Burbank, published an exhaustive study of the Ricardo group of formations. Using as reference points the ages of lava flows and volcanic ash beds within the sedimentary sequence, they created a detailed time scale for all parts of the Ricardo group of formations. They estimated rates of sinking of El Paso basin and the rate of deposition within it, both of which varied with time. They also measured ancient changes in the earth's magnetic field as recorded by magnetic minerals within sedimentary layers of the Dove Spring formation. They obtained ages of the rock layers by matching the local magnetic record against a master chart compiled from worldwide analyses of seafloor sediments, where the full magnetic record is preserved.

The data of Loomis and Burbank show that accumulation of the Cudahy Camp formation started about 19 million years ago and continued for 6 million years, when it was interrupted for about 1.5 million years. During this interruption, erosion created the surface on which the Dove Spring formation was deposited. The Dove Spring formation accumulated from 11.5 to something less than 7 million years ago.

Loomis and Burbank used just about every known geological technique to determine the sources of the sediment dumped into the slowly subsiding El Paso basin. The material in the Cudahy Camp formation came primarily from the Mojave block immediately to the south. It also includes sediment eroded from the underlying Goler formation, as well as from 300-million-year-old Permian sedimentary rocks and the 470-million-year-old Ordovician Mesquite schist. Much of the debris eroded from the Mojave block is volcanic ash and fragments from dark lava flows and other volcanic rocks. The sources of those volcanic rocks are now gone, mostly because that terrane has moved eastward along the Garlock fault. Earlier vertical movement certainly occurred along the Garlock fault, but the Dove Spring beds tell us that horizontal movement started only ten million years ago. Ten million years of offset along the Garlock fault, estimated at 0.16 to 0.25 inch per year, means that the source areas for sediments in the Ricardo group now lie far to the east.

Sediment in the Dove Spring formation also came, initially, from the Mojave block. Although largely volcanic, the variety of source rocks was much greater than in the Cudahy Camp formation. More pieces of relatively old granite and metamorphic rocks suggest that erosion had stripped the volcanic cover from some areas, exposing the older rocks beneath. Ash beds lie between layers of sediment throughout the Dove Spring formation, especially near its base, the rocks in the colorful cliffs alongside California 14.

About eight million years ago, the proportion of granitic debris in the Dove Spring formation increased dramatically, and its character changed from that of the granite in the Mojave Desert to that of the Sierra Nevada granites. Evidently, at that time, uplift along the southern part of the Sierra boundary fault began to raise the Sierra Nevada, an event geologists have wanted to date for a long time. It seems that the southernmost Sierra Nevada began to rise something less than eight million years ago. They still shed sediment into and over the Red Rock Canyon area today.

VIGNETTE 13

A GREAT NATURAL GEOLOGICAL LABORATORY

Rainbow Basin

SAN BERNARDINO COUNTY

A canyon famous for its fossils slices through the Mojave Desert a few miles north of Barstow. Rocks exposed in its walls contain the bones of a rich assortment of bears, giant pigs, camels, and other animals that roamed the land and sipped water from a shallow lake about 15 million years ago. The canyon is in a small range known by the unglamorous name of Mud Hills. It contains Rainbow Basin, a scenic and geologic gem.

We can think of three good reasons to pay Rainbow Basin a visit. First, it is a wonderfully colorful spot where the layered sedimentary rocks crop out in shades of bright green, white, red, and brown—hence the name of the basin. Second, it is a key link in our knowledge of the mammals that lived during middle Miocene time. Paleontologists named an interval of geologic time, the Barstovian stage (12 to 16 million years ago), after the fossils found here. Third, the rocks in Rainbow Basin are folded and faulted in a manner that mimics in miniature the complex structures in the Transverse Ranges of southern California. For these reasons, Rainbow Basin was declared a National Natural Landmark in 1972.

In this vignette, we describe two excursions in the Mud Hills: a short drive through Rainbow Basin to see what is visible from the car, without hiking; and a hike up a nearby canyon, which takes a half day and requires a few scrambles over small dry waterfalls. Rock and fossil collecting is prohibited within the landmark area. If you find any interesting fossils,

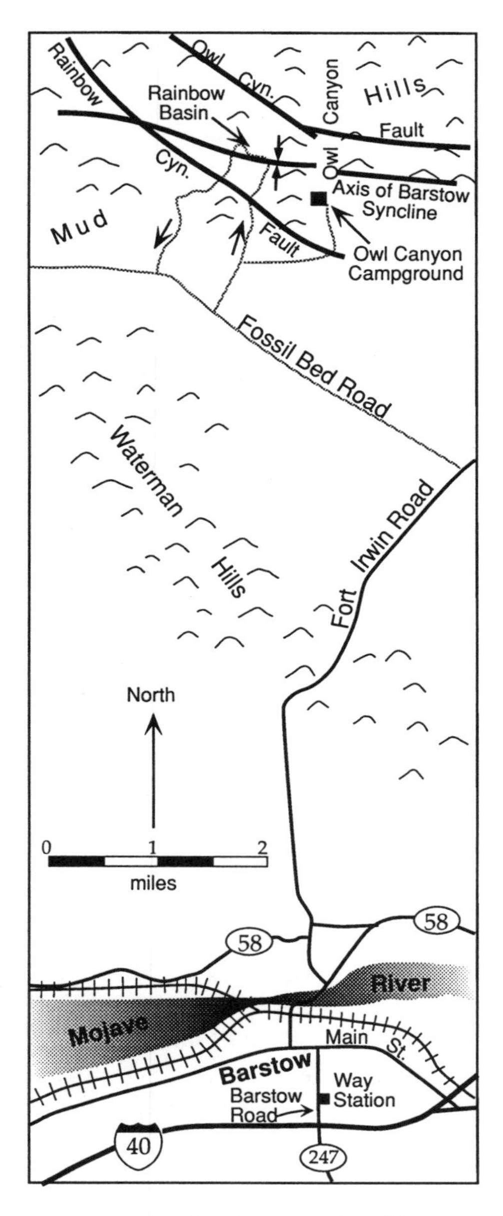

Access to Rainbow Basin and Owl Canyon. Barstow Way Station, a Bureau of Land Management facility, is a good place to begin your trip.

The auto club's San Bernardino County map and the United States Geological Survey, 7.5-minute, Mud Hills topographic quadrangle cover the area. People coming from the Los Angeles area should take the Barstow Road exit (California 247) north from Interstate 15. Two-tenths of a mile north of the freeway is the Barstow Way Station (831 Barstow Road), a Bureau of Land Management facility that houses exhibits about the high-desert environment and provides information about road conditions. The Way Station also houses the Old Woman meteorite, a 2.75-ton chunk of nickel and iron that was found in 1976 in the Old Woman Mountains, about 100 miles east of Barstow.

Eight-tenths of a mile north of the freeway, Barstow Road ends at a T-intersection with Main Street. Turn west (left) and go 0.2 mile to First Avenue. Turn north (right) on First Avenue (follow the signs to California 58, Bakersfield, Fort Irwin, Las Vegas); cross the Mojave River and railroad yard, and then curve east (right) around the Harvey House train station. Note the extensive railroad facilities; Barstow was founded in 1886 as a railroad town, and for many decades, the railroad was the main employer there. In fact, Barstow was named for William Barstow Strong, then president of the Santa Fe Railroad. (Strong's middle name was chosen because his last name had already been used for a Santa Fe station in Kansas.)

Nine-tenths of a mile from the last curve, turn left onto Fort Irwin Road (to California 58, Hinkley). This turn is just past the small hill capped by a white water tank. Proceed north on Fort Irwin Road, passing through two stop signs (after stopping, of course) at California 58. Travelers coming to Barstow from Mojave or Bakersfield via California 58 join the route at the first stop sign. Five and a half miles north from the second stop sign, a well-graded, labeled, wide gravel road bears off to the left; take it. This road (Fossil Bed Road) is marked on the right by a sign labeled "Rainbow Basin". Drive three miles and then turn right to the entrance to Rainbow Basin.

please contact either the San Bernardino or Los Angeles County museums.

The best way to begin your visit is to drive the one-way loop through the basin. Continue north on the access road, past the turnoff to Owl Canyon. The road shortly ascends a tight little canyon that floods during storms. Notice that the rock layers there tilt down to the north (ahead). After several tight twists and turns, the road curves west into an open basin.

Notice that all the rocks on the far side of the basin are inclined toward you; they tilt down to the south. The reason for this soon becomes clear. Drive west a few hundred yards to where the road climbs a small hill and comes to a parking area on the north (right). Walk east about twenty yards to a saddle between two knobs. This is a good place to look at the Barstow beds.

Observe that all rocks on the north wall of the basin dip south, and all those on the south wall, through which you just came, dip north. The two walls are on opposite flanks of a broad synclinal trough, which is clearly visible in the east wall of the basin. This is the Barstow syncline, a fold so obvious and beautifully displayed that it is a geological celebrity, its photograph is featured in many textbooks.

The Barstow syncline is like a wrinkle in a rug, formed as crustal forces compressed the sedimentary layers. The trough of the fold trends from

Barstow syncline, looking east from the parking area in Rainbow Basin. Skyline is capped by flat-lying alluvium that lies unconformably on tilted Barstow beds. The flat surface in the middle foreground is a low-lying version of this same unconformity.

east to west, so the compressive forces must have been applied from north and south. The Mud Hills lie athwart the Calico fault, a major structure that reaches from the eastern San Bernardino Mountains northwest across the Mojave Desert almost to El Paso Mountains. The Calico fault has the same trend and moves in the same way as the notorious San Andreas fault—as do a swarm of others, including the Pisgah, Helendale, and Camp Rock faults. The Camp Rock fault broke in the massive Landers earthquake of June 1992. All move horizontally such that if you stand on either side of the fault, you see the opposite side moving to your right.

In a few places, notably the Mud Hills, the Calico fault bends. During a displacement, rocks on opposite sides of the fault press into each other at the bend and therefore buckle into folds. The Barstow syncline formed because the Calico fault bends to the west. The situation resembles the Big Bend in the San Andreas fault.

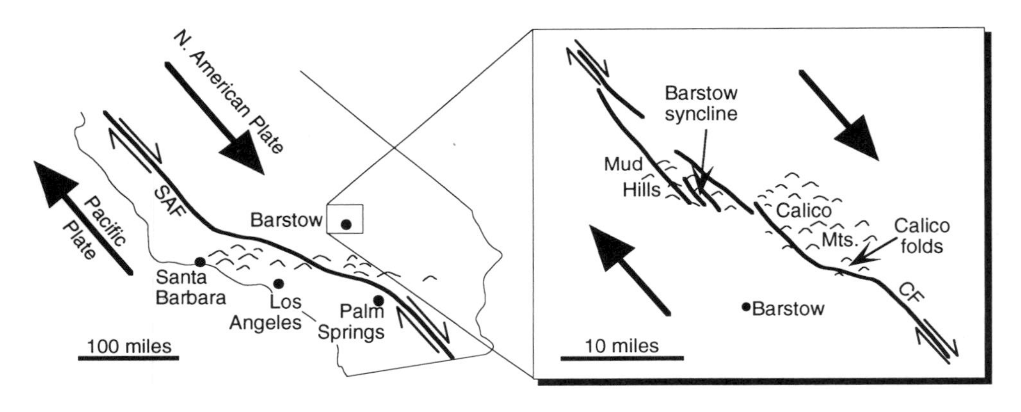

Bends in strike-slip faults can cause folds, thrust faults, and earthquakes. North of Los Angeles, a westward bend in the San Andreas fault produced the Transverse Ranges. North of Barstow, a westward bend in the Calico fault system has produced spectacular folds in the Mud Hills and Calico Mountains.

South of Palm Springs, the San Andreas fault trends about 50 degrees west of north. North of that resort town, the fault assumes a more westerly trend (about 60 degrees west of north). Near the southwest end of the San Joaquin Valley, north of Los Angeles, the fault again trends more northerly (about 40 degrees west of north). The fault segment between Palm Springs and the San Joaquin Valley is known as the Big Bend.

All of California west of the San Andreas fault is part of the Pacific plate, moving northwest relative to the North American plate, which contains the rest of California and the contiguous United States. Give the San Andreas fault another 10 or 20 million years and it will move coastal southern California next to San Francisco, which lies east of the fault—thus providing rich food for political and cultural thought. As the Pacific plate and its passengers move northwest, parts of California near the Big Bend, specifically the east-west Transverse Ranges, rise because the plates push into each other. The results of that encounter include mountains, folds, and earthquakes (such as the Kern County earthquake of 1952 and the San Fernando earthquake of 1971, both of which were triggered by compressive forces in the earth's crust).

Several branches of the Calico fault are visible from the parking area. One cuts the east wall of the basin, just north of the synclinal trough. This fault, the Owl Canyon branch of the Calico fault, is easy to spot because it chops off the sedimentary layers in the syncline's southern limb and places them against layers of a different color. Turn around and look at the west wall of the basin, and you will see one or two other faults. A close look will reveal that many small faults cut the rocks throughout the basin. The faults show a few to tens of feet of displacement and trend northwest. These little faults are parts of the mechanism by which the Mud Hills yield to prevailing crustal stresses.

If you look from the parking area to the east wall of the basin, you can see an angular unconformity between the tilted rocks and a thin cap of sand and gravel with Joshua trees growing on it. Geologists adore angular unconformities because they record so much geologic history, especially tectonic events. This one records the folding of the lower, 15 million year old, layered rocks, their truncation by erosion, and then deposition of the sand and gravel above. The sand and gravel cap is no more than a million years old, so the buried erosional surface of the unconformity represents

Tilted beds on the north side of Rainbow Basin. The Skyline tuff is the prominent white layer. If you look at the tuff from this spot, the layers dip toward you, pass underneath, and reemerge on the south side of the syncline, behind you.

about 14 million years for which no sedimentary record exists at this place.

The Barstow formation in Rainbow Basin consists dominantly of mudstone and sandstone that were deposited in a lake. The several layers of extremely white rock are volcanic ash of rhyolitic composition. Isotopic dating, utilizing the decay of potassium to argon, of the ash shows that they are between about 18 and 13 million years old, so that must also be the age of the Barstow formation, of which the ash layers are part. The formation was deposited long after the dinosaurs vanished 65 million years ago and long before the first hints of humanity appeared in Africa, perhaps 3 million years ago.

Several volcanic ash beds are visible from the parking area. Most conspicuous is the Skyline tuff, a white or buff layer about ten feet thick on the narrow skyline ridge northwest of the parking area. Volcanic ash, when it becomes consolidated to form rock, is known as tuff (not to be confused with tufa, which is calcium carbonate deposited from the water of springs or streams). The tuff bed dips south, under your feet, and emerges again on the south limb of the syncline where it dips northward and is harder to spot from here. As you drive west from the parking area, the road passes close to the Skyline tuff just before swinging sharply south. Here the Skyline tuff on the north limb of the syncline, just north of the road, is folded into another, much smaller syncline. Another prominent white tuff is exposed in a bottleneck passage after the road exits the basin. Mudcrack impressions on the tuff's lower surface indicate that the lake in the basin dried up on occasion.

A secondary syncline on the north limb of the Barstow syncline, north side of Rainbow Basin. Here, the Skyline tuff and surrounding beds are folded into a broad syncline and cut by several northeast-trending faults. One fault has dropped a piece of the Skyline tuff down to the center of the photo.

Cavernous weathering in the Barstow formation creates shallow caves that are fun to explore. Clay minerals in these rocks expand when wet, producing a slippery popcorn-like surface that coats exposures in the aptly named Mud Hills.

Many of the rocks in the Barstow formation are bright green. Novice geology students tend to assume that green rocks indicate the presence of copper, a potent rock dye. However, most of the green in these rocks comes from clay minerals that form during weathering. Volcanic glasses readily alter to slippery green and brown clay.

Paleontologists found fossil bones in Rainbow Basin in the first two decades of the twentieth century and have since combed the area so thoroughly that few remain on the surface. But fossils do weather out of

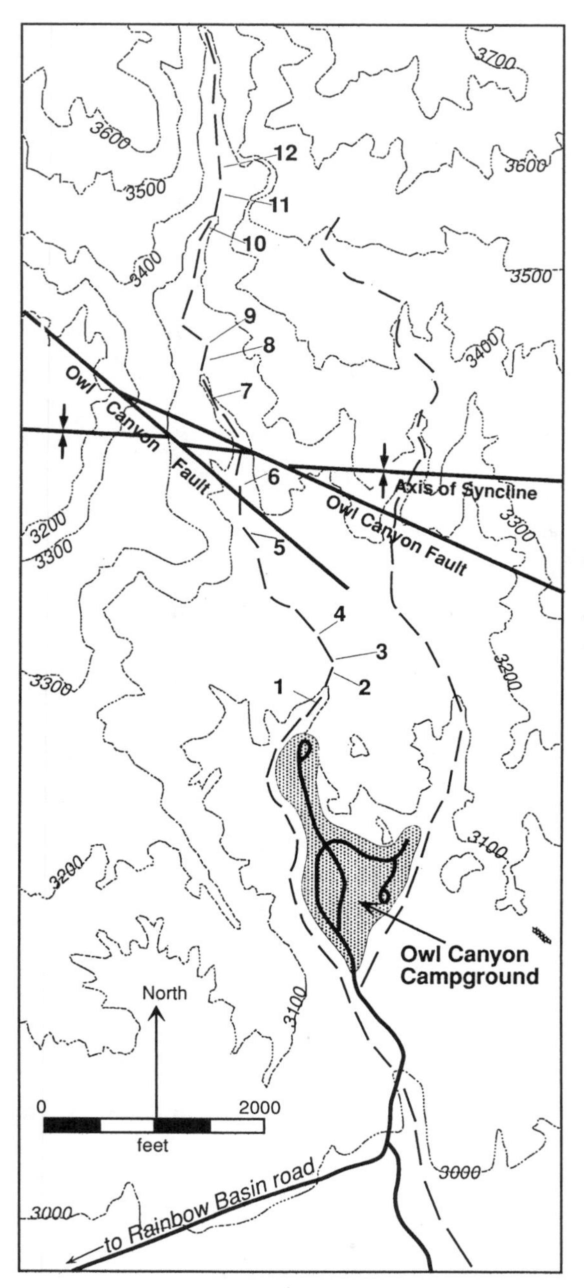

Topographic map of lower part of Owl Canyon hike. Contour interval is 100 feet. Numbers give locations of features discussed in text.

the rock, and every hard rain reveals a few. The remains are remarkably diverse. They include bones, teeth, and tracks of primitive bears, dogs, antelopes, rhinos, pigs, horses, camels, mastodons, rodents, and extinct animals called oreodonts that resembled pigs. Insect and plant remains, including oak, poison oak, palm, and juniper, are also present. They suggest that the climate was similar to that of coastal southern California today, probably with 15 to 20 inches of rainfall. John C. Merriam, the paleontologist who first studied this area and named the Barstow formation, concluded from the abundant and diverse fossil remains that the climate at that time was warm, with alternating dry and wet seasons. The area was a plain, much like the African savannas, where large herds grazed on the lush growth of grasses and foliage.

To examine the rocks of the Mud Hills in greater detail, walk up Owl Canyon, the first major canyon east of Rainbow Basin. To reach the trailhead, drive to the Owl Canyon campground, then north to the north end of the camping area. Walk to the bottom of the wash and follow it, avoiding any trails leading out of the wash. Bring a flashlight. Numbers in parentheses in the following text are keyed to the location map of Owl Canyon (page 118).

Before starting your hike, look at the north wall of the gully. Pink gravels lie on beds of the Barstow formation that dip about 40 degrees upstream. This is the same angular unconformity that you saw in Rainbow Basin, and you will see it many more times in Owl Canyon. Notice that the layers of the Barstow formation tilt down to the north, upstream, as in the southern part of Rainbow Basin. This means that you are on the south side of the syncline. Rocks here are mostly mudstone, siltstone, and sandstone with occasional beds of pale volcanic ash. All were deposited in lakes or along streams.

About 100 yards from the trailhead, at a sharp zig-zag bend (1), horizontal layers of young gravels are plastered onto the canyon wall at about eye height. They are just a coating left from a former gravel filling of the canyon, which has been filled and sluiced countless times in its history. About 200 feet farther, the stream makes a second zig-zag (2). On the east wall here, you will find a lovely exposure of the unconformity, where pink gravel is piled up against bedding planes of the Barstow formation.

Another 100 feet or so upstream (3), again on the east side, a little hillock of green beds in the Barstow formation is almost completely covered by younger pink gravels. Nearby is a huge (seven-foot) boulder of tuff breccia derived from the underlying Pickhandle formation, which crops out upstream. We will see more of this formation later. About 50 feet beyond the boulder, a side trail takes off up the east bank. Keep to the gully floor.

Unconformity between young gravels and Barstow beds in lower Canyon (stop 2 of hike). Here, horizontally bedded young gravels (left) were deposited on bedding planes of the Barstow formation (right).

About 200 feet beyond the side trail, a hard, white bed, about one foot thick, crops out on both sides of the wash (4). This is a layer of tuff, produced by an ancient volcanic eruption. Like all the rocks you have walked through so far, it dips to the north, upstream; thus, you have been walking into younger and younger rocks. A few hundred feet farther, bedding in the rocks flattens out and the inclination begins to shift back to the south (5). This change is complicated by several small faults that cut through the canyon at this point. Although not wholly obvious, this spot marks the axis of the Barstow syncline. As you walk north from here, you pass into older and older rocks.

Shortly after you cross the axis of the syncline, you come to what appears to be a major fork in the canyon (6). The right fork is the easier route, but either will work because this is not really a fork. The left "fork" cut off a loop in the channel. The two channels join a short distance upstream.

Within a short distance, the canyon wall becomes higher and more dissected. After a few hundred feet between close valley walls, you reach a section where the stream bank collapsed (7), making the channel wider.

A little farther, large boulders of sandstone from the Barstow formation have tumbled into the canyon, blocking the channel (8).

About 150 feet beyond the fallen blocks, watch for the entrance to a natural tunnel in the east wall as you round a bend (9). You will need your flashlight to explore it. This tunnel, which is about twenty yards long, links Owl Canyon to a small tributary canyon to the east. During heavy rains, water flows through the tunnel from the tributary canyon to Owl

Undercut stream bank in Owl Canyon, near stop 7 of hike. Collapse of cuts like this widens the stream.

Entrance to unusual tunnel on east side of Owl Canyon, stop 9 of hike. Bring a flashlight and explore.

Canyon. Natural tunnels are quite unusual. This one probably formed as water trickled through cracks in the sandstone.

For the next several hundred yards, the rocks are mostly sandy conglomerate, a pebble-rich rock. After a major bend a few hundred feet beyond the tunnel, the wash narrows impressively, with steep walls closing in on either side. Eventually, near the base of the Barstow formation, you see beds of coarse fanglomerate, a conglomerate formed from an alluvial fan. Examine the cobbles in the fanglomerate. They are granite made of large crystals (¼ inch or so) of quartz, feldspar, and mica. These cobbles are identical to the granite exposed in the northern part of the Mud Hills. Clearly, the fanglomerate records a depositional environment quite different from that in which the finer mudstones were laid down. Our walk has taken us back in time, from a quiet lake to a more active alluvial-fan environment in which torrential rains brought boulder-laden debris flows into the basin.

After about 500 feet of this narrow section, a fault crosses the wash bottom (10). The fault dips about 80 degrees downstream (south) and has

Boulders of Pickhandle breccia on floor of Owl Canyon, about 200 yards downstream from dry waterfall. Staff is about 4 feet long.

Narrow upper part of Owl Canyon, cut into Pickhandle breccias above the dry waterfall. Canyon here is about 8 feet wide.

brought finer, softer sandstone and mudstone into view. About 70 feet upstream of the fault is a huge (ten-foot) boulder of brown breccia made of sharply angular rock fragments from the Pickhandle formation. Some of the pieces are more than a foot across. All are granite. Feel free to pat the boulder.

A few hundred feet farther, the wash walls become very high and vertical and seem to hang over you. Near the top they show cavernous weathering on a small scale. Soon the easy walk ends at a dry waterfall (11) carved out of the more resistant rocks of the Pickhandle formation. It marks the contact between the Pickhandle and Barstow formations. You have walked completely through the Barstow formation. The contact between the Barstow and Pickhandle formations is another angular unconformity. The layers of rock in the Pickhandle formation are here inclined about 20 degrees steeper than those in the overlying Barstow formation.

From this point north, the canyon walls become much steeper and the rocks are composed of much coarser fragments. Gone are the smooth, rounded pebbles and cobbles of the conglomerate. The Pickhandle formation is a stack of breccias and volcanic mudflows. The Pickhandle breccias presumably tumbled from steep, cliff-like fault scarps in granite.

Your walk up the canyon has taken you back in time from the mudstone of the upper Barstow formation, which was deposited in quiet lake water, into alluvial fan gravels of the lower Barstow formation, and finally into the fault scarp breccias of the underlying Pickhandle formation. The layered rocks of Owl Canyon record a time when mountains rose along a fault, eroded, and eventually were submerged in a lake.

Farther north, the canyon remains narrow and becomes impressively deep. In its deepest parts the walls are made of large, shattered blocks of granite interbedded with sedimentary layers. These blocks slid off the fault scarps and into the accumulating Pickhandle formation. They are smaller versions of landslides, such as the Blackhawk Slide (Vignette 16), and tell of vigorous faulting and mountain building in this area.

At its northern end, the canyon opens into a small basin as the breccias and mudflows give way to more easily eroded conglomerate at the base of the sedimentary sequence of the Mud Hills. If you walk a little farther north you will find that Owl Canyon heads on the slopes of a granite ridge that rises to the crest of the Mud Hills. These granite slopes, with their Joshua trees and rounded outcrops, make an excellent place to have lunch, and to ponder the geological history of the Mud Hills.

VIGNETTE 14

YOUNG VOLCANOES IN THE MOJAVE DESERT

Walking About at Amboy and Pisgah Craters

SAN BERNARDINO COUNTY

Relatively young volcanoes dot the triangular region of the Mojave Desert between Barstow, Baker, and Twentynine Palms. Amboy and Pisgah craters are among the youngest of those youthful volcanoes. Both have small craters in their summits, but mainly they are excellent examples of cinder cones. Cinder cones are small basaltic volcanoes that spit spongy fragments of basalt lava up to an inch or two in diameter, the sort of stuff you might buy in bags to spread beneath your shrubs. Geologists call such material tephra, a term Aristotle used that is based on the Greek word for ashes. Tephra refers to any fragmental material (ash, cinders, or even automobile-sized blocks) that a volcano blows into the air.

Cinder cones typically appear, erupt tephra for a few weeks or years, produce lava flows, then stop. Few ever erupt again. In most cases, the next eruption will build a new cinder cone. The larger tephra blown from a cinder cone piles up around the vent to build the volcano; the smaller material drifts downwind as a cloud of ash. Although a strong wind may shape them, most cinder cones are symmetrical. After the column of magma has blown off most of its gases, the tephra eruption ceases, and the volcano ends its career by producing a lava flow, perhaps two or three. The flows typically burst out of the base of the cone because the pile of loose tephra is too weak to support an internal column of magma. In many cases, a segment of the cone floats away on a lava flow. The western side of Amboy Crater has just such a breach.

Parícutin, west of Mexico City, may be the world's most famous cinder cone. It unexpectedly appeared in 1943 in a cornfield, then grew to a height of 1,200 feet during the next nine years, erupting lava flows that covered more than ten square miles and burying a village. It is a giant among cinder cones. Although the sudden appearance of Parícutin certainly startled the farmer who watched the eruption begin in his cornfield, it should not have been much of a surprise. The surrounding area contains about 200 cinder cones.

Amboy and Pisgah craters stand amid large fields of basalt lava. Although the tephra eruptions undoubtedly made good fireworks displays, they were neither dangerous nor violent—by volcanic standards. Basalt flows typically erupt very quietly and predictably. Contrast that to rhyolite eruptions, such as that of the Bishop tuff from the Long Valley caldera (Vignette 19), which may devastate a region the size of an average state, and possibly change the climate worldwide for years.

Most of the basalt lava flows erupted from Amboy and Pisgah craters have "pahoehoe" surfaces. Pahoehoe, a Hawaiian term, refers to surfaces that are relatively smooth and locally covered with distinctive ropy coils. This ropy texture forms when molten lava beneath rumples a thin outer skin of hardening rock, much like the surface texture of fudge that cools as it is poured into a pan. Some of the flows have "aa" surfaces, another Hawaiian term. Aa surfaces are rough, clinkery, and jagged; they tear your boots.

Broken, tilted block of pahoehoe near Amboy Crater.

A narrow patch of pahoehoe makes the only smooth trail among jagged aa flows near Pisgah Crater.

Although they look quite different, flows with pahoehoe and aa surfaces are actually the same kind of rock; they have the same composition. The difference in flow surfaces depends upon how much steam the molten lava contains. Steam makes basalt lava more fluid, and therefore likely to develop a pahoehoe surface; dry lava is considerably more viscous, and tends to develop an aa surface. Indeed, both surface textures may form on the same flow; a smooth pahoehoe surface near the vent may fade into a rough aa surface farther away, reflecting the flow's loss of steam to the air.

How old are Amboy and Pisgah craters? No one knows for sure, except that they are among the youngest volcanoes in southern California. They are too young to date by the potassium-argon technique, which can be used on basalt erupted as recently as about 100,000 years ago. They may be young enough to date by the carbon-14 technique, which can be used on samples as old as about 50,000 years. But geologists normally use volcanically charred wood for radiocarbon (carbon-14) dating, and so far no one has found any around either cone. Neither volcano shows much evidence of erosion, but that is typical of cinder cones; the loose tephra absorbs water so well that surface runoff rarely gullies the surface. Lavas from Amboy Crater are interbedded at a depth of about 30 feet with lake sediments in Bristol playa (Vignette 15) that are about 100,000 years old.

So these volcanoes could be as much as 100,000 years old, perhaps much younger.

The best way to enjoy Amboy and Pisgah craters is to hike. Be sure to bring plenty of water; black basalt really bakes in the sun. Hiking is easy, except across aa surfaces or deep fissures. The weird lava shapes, the interplay between lava and the wind driven sand, and the creatures that live in the rocks and leave their tracks on the sand are all worth seeing.

Amboy Crater

At Amboy Crater you can see lava flows and an absolutely undisturbed cinder cone. It is just southwest of Amboy, which was a lively settlement in the days when U.S. 66 was a major artery. (Amboy, by the way, marks the west end of an alphabetically named series of way stations along the Santa Fe Railroad—Amboy, Bristol, Cadiz, Danby, and so on.)

Bristol Dry Lake, just east of Amboy Crater, is a desiccated remnant of a much larger lake that flooded this undrained desert valley during the ice ages, when the climate was much wetter than now. It contains salts, chiefly sodium and calcium chloride, that weathered out of the rocks in the surrounding mountains. They accumulate on the desert floor because the basin has no outlet stream, so the salts become concentrated by evaporation of whatever water collects there. A sort of mining operation harvests the dissolved salts through a system of ditches that drain brine from the sediments in the valley floor and carry it to large evaporating flats. Watch for the white crusts of salt on the sides of the ditches.

Locations of Pisgah and Amboy craters.

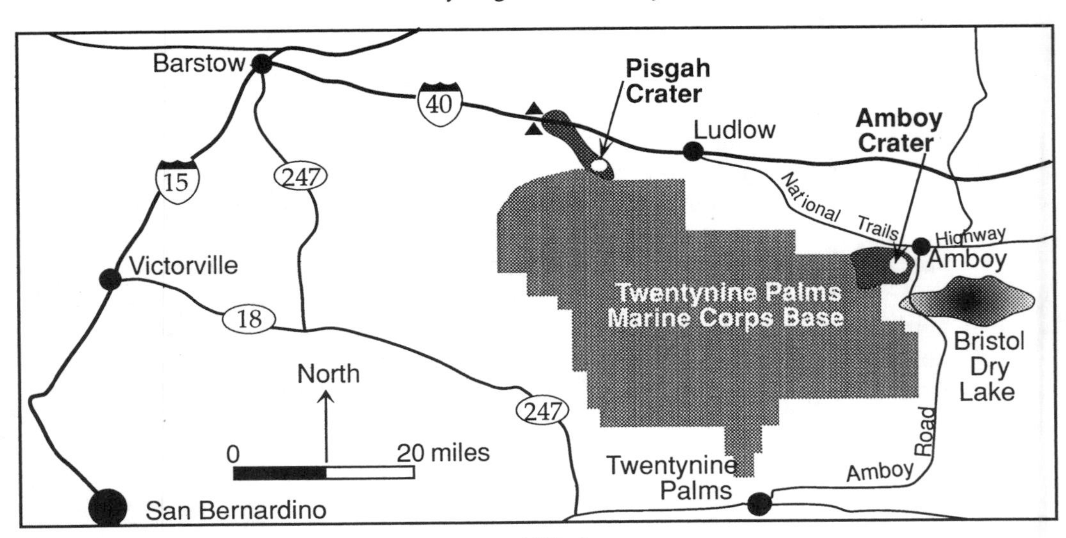

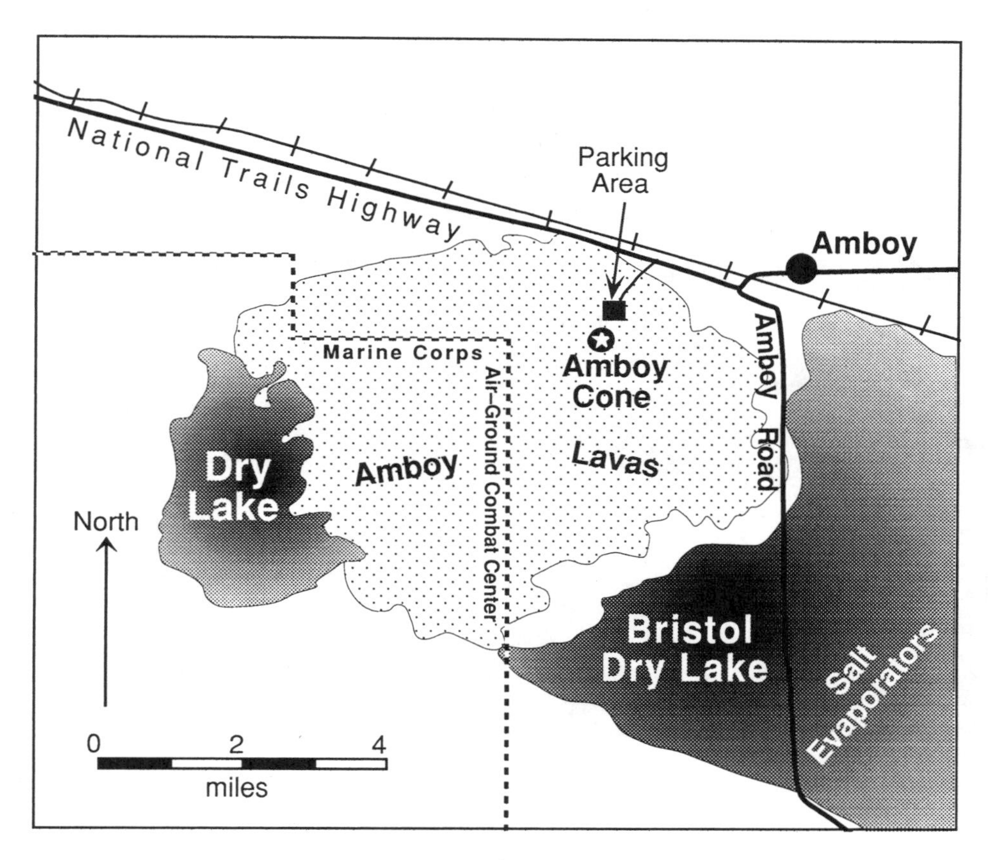

Access to Amboy Crater.

Amboy Crater is easy to reach from Barstow or Twentynine Palms. The U.S. Geological Survey, 7.5-minute, Amboy Crater topographic quadrangle covers the area.

From Barstow, drive 52 miles east on Interstate 40 to Ludlow. Turn south (right) on the off-ramp, then immediately east (left) at the stop sign. Drive 26 miles east on National Trails Highway to an unmarked turnoff south onto a graded dirt road. This turnoff is 1.3 miles east of the bridge where the highway first crosses onto lava. It is one mile west of the intersection of Amboy Road and the National Trails Highway.

If you're coming from Twentynine Palms, take Amboy Road east and north 49 miles. You will cross onto the playa of Bristol Dry Lake about nine miles south of Amboy. Continue to the National Trails Highway and turn west (left). Drive one mile west to a turnoff on the left onto a graded dirt road.

The easiest way to reach the cinder cone is to drive 0.4 mile down the short graded road that heads south from the National Trails Highway to a parking area at the edge of the first good lava exposures. Then you can wander short distances into the lavas south or west of the parking area.

Aerial view, looking northeast, of Amboy cone, with National Trails Highway and settlement of Amboy in the distance. Lavas are covered by a light-colored mantle of eolian sand which is absent downwind (southeast) of the cone. Note breach in cone and inner conelets. —John S. Shelton photo (taken in 1957)

Walking is easiest on the low ground, on the sand and volcanic rubble. Be careful to check for landmarks and directions, lest you lose your way in the chaotic lava surface. Plan to follow your own footsteps back or to intersect the track that runs south from the parking area. A compass may prove useful, but beware of erroneous readings; the high iron content of the lava flows can distort the local magnetic field. Using a distant peak as a reference direction is also helpful.

Look for remnants of ropy pahoehoe surfaces. Many survive on steeply inclined faces, obviously tilted as the molten lava within the flow continued to move after the surface formed. In some places, the moving lava raised parts of the surface; in other places, the surface dropped as lava flowed out from beneath it. Watch for occasional slabs of rock with patterns of parallel grooves that formed when pasty lava squeezed like toothpaste through a jagged crack.

Many of the lava surfaces display cracks that define four-, five-, or six-sided polygons a foot or so across (see Vignette 19). These are the tops of the palisades of vertical columns you may see in basalt flows exposed in ledges or road cuts. They form as the lava shrinks slightly when it crystallizes, but while the solid lava is still very hot.

As you explore the lava flows, watch for pressure ridges—great slabs of lava that buckled up due to pressure from still-liquid lava flowing beneath the congealing crust. These linear ridges, with open cracks along their crests, form most commonly in pahoehoe flows. Watch for places where molten lava squeezed up through the crack in the crest of the pressure ridge to make great blobs that resemble black taffy. These bulbous extrusions are called "squeeze-ups." In other places, the surface of the flow collapsed to form depressions as the molten lava drained out from beneath it.

Most of the lava on the flow surface is full of gas pockets and bubbles left where steam did not quite escape. Although the top of the flow is relatively smooth overall, it is not really a very good example of a pahoehoe surface because it lacks the typical ropy texture.

The cone is about a mile from the parking area, and its circumference is another mile, so the round trip is roughly three miles. Allow three hours, minimum. The whole trek is a dandy adventure, a miniature expedition. Don't forget to carry plenty of water.

An easy walk brings you to the base of Amboy Crater in less than thirty minutes. Follow the fading vehicle track about a half mile until it branches in several directions, then continue along one of the foot trails to the cone. Instead of climbing 220 feet up the steep trail on the north side of the cone, follow the vehicle track counterclockwise around the base.

Saline pool in excavated ditch, Bristol Dry Lake, just off Amboy Road.

A trail through the breach in the west wall climbs only 80 feet to the floor of the inner crater. Another trail ascends a gentle slope from inside the crater to its rim, where you get a good view. The loose rubble on the cone makes it hard to descend any slope without slipping.

A history of an eruption in several phases can be read from the shape of the cone. The main edifice formed first, during the main period of tephra eruption. A lava flow then breached the west side of the cone. You can see a bit of that flow within the breach. Two minor explosive pulses followed, producing two small nested tephra rings inside the main crater.

Leave the crater through the breach in its west wall, and continue the walk around its base. At the south side is a large depression with a flat floor about 40 feet below the level of the track. You see it best from a small bench at the base of the cone. Slopes above the bench display scars of a landslide, which probably created the bench. You can reach the floor of the flat by way of one of the gullies in the bench.

Along the way, you can see why Amboy Crater has escaped cinder mining. Its debris is too coarse. Much consists of solid lava chunks, 20 inches or so in diameter, and of even larger masses of agglutinated lava blobs, which stuck together while still partly molten. Commercial operators prefer very bubbly cinders in the pea to walnut size range. Such material is scarce in Amboy Crater, a deficiency to which Amboy owes its preservation, in contrast to Pisgah.

Upon crossing the flat floor of the depression at the south base of the cone, you will come to a lava flow of shiny, black basalt with a convoluted pahoehoe surface. Look for a band of lava close to the base of the cone that is free of rubble, somewhat smoothed and worn, and a little lighter in color. A gully more than three feet deep runs from the base of this band across the flat floor of the depression. Its appearance suggests that quite a lot of water has flowed down the face of this band. It is a dry cascade. If you step up the 20 feet of blocky wall to its top, you will see a dry stream bed of clean rock curving out of sight around the east side of the cone. Where did this water came from?

About opposite the dry cascade, the slopes of Amboy Crater change from large chunks of lava to small rectangular chunks of surprisingly uniform size, between two and five inches across, along with a goodly sprinkling of spindle-shaped volcanic bombs. Rills and gullies dissect these slopes, which may be an older part of Amboy Crater. Alternatively, the eastern side of the cone may be gullied because it is on the downwind side. Silt blown over the cone may be trapped there among the cinders. This silt would then prevent water from passing through the cone, causing it instead to course down the surface.

In another 200 or so yards, you come to a larger and deeper gully, with tributaries. The stream bed ends here. Look up the slope to see a number of pipelike openings on the walls and at the heads of the gullies. One has

Cavern at the mouth of fluvial pipe that channeled water from inside Amboy Crater. Exposure is a wall of stream-cut gully fed by pipes on the east side of Amboy cone. Note the layering within cobble-size lava fragments.

a diameter of at least three feet. These look like pipes coming out of the cone, and indeed they are natural pipes, although their actual diameters are much smaller than the cavernous openings you see. Water flowing from the crater through these natural conduits eroded the slope gullies, flowed along the stream course, poured over the dry cascade, and dissected the floor of the depression. The gully walls show nicely the layers of small uniform blocks that form the wall of the cone.

Rilled slopes, but without pipes and gullies, continue on the northeast side of the cone. You can see a distinct contact between the rilled older slopes and the smooth younger slopes. It appears that more than half of an older cone had been destroyed before the younger cone covered most of it. The contact between the two, which dips down to the west at an angle of about 45 degrees, becomes more obvious if you look back from the broad flat north of the cone.

Your car, in the parking area, should come into sight to the north as you round the northeast corner of the cone. You can set a straight course for it across relatively easy terrain, but select a couple of distinctive landmarks well up toward and on the skyline of the far valley wall. Your car will drop out of view again, and these navigational aids will help you keep on course until you can see your car again.

Pisgah Crater

People who do not want to make the long trek to Amboy Crater may wish to visit Pisgah Crater, which is closer to Los Angeles. But cinder quarries put some parts of Pisgah Crater off limits.

A series of stops along the quarry haulage road provides good views of the lava flows. The first is at a wide place about 0.2 to 0.4 mile south of the highway. Walk east 200 to 300 yards across lava largely covered with sand. Without its blanket of windblown sand and silt, the surface of this flow would be fairly rough.

Now drive to a point 0.6 mile south of National Trails Highway and walk east. Sand buries only about 15 to 20 percent of these flows, which lie upon the flows of the previous stop. You might suspect that this difference in cover reflects a difference in age. The rocks' magnetism, though, indicates that the three main sets of lava flows at Pisgah Crater, including these, erupted within a few decades of each other. The difference in cover, then, probably reflects a difference in sand supply. When you return to your car, look back at the abrupt front edge of the younger flows rising seven to ten feet higher than the flows at the first stop.

Pisgah cone, viewed from the west. Dust is from ongoing quarry operations, which have cut deeply into the cone on the north. White rectangle is large trailer at quarry site. Aa in foreground.

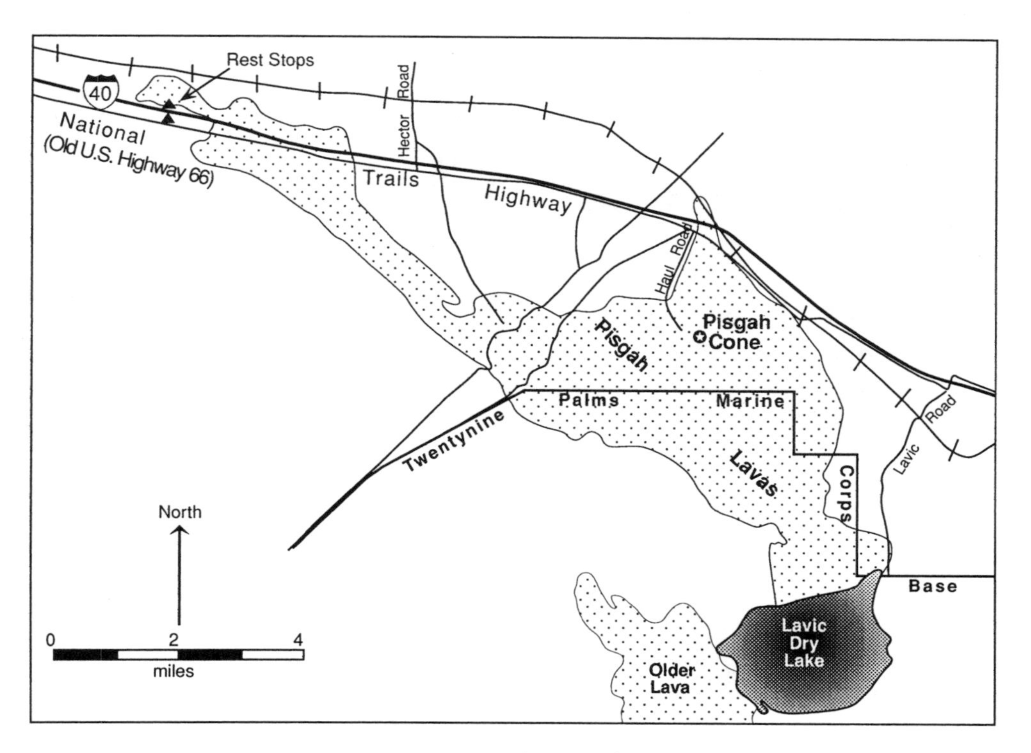

Access to Pisgah Crater lavas.

The U.S. Geological Survey, 7.5-minute, Sunshine Peak, Lavic Lake, and Hector topographic quadrangles cover the area. Pisgah Crater is 40 miles east of Barstow, just south of Interstate 40. The rest stop about 30 miles east of Barstow is just beyond the west edge of the Pisgah lavas. Pisgah Crater is the skyline knob in the valley ahead. The freeway crosses lavas for about two miles, dropping off at the sign that announces "Hector Road Exit, 1 mile." These lava flows traveled a full ten miles from the cone. At Hector Road exit, 34.5 miles from Barstow, turn south (right) and then east (left) at the T-intersection with National Trails Highway (old U.S. 66). Drive east 4.6 miles to the haul road for the cinder quarry. It provides access to the lava flows around the cone.

Breached pressure ridge near Pisgah cone.

While walking on these flows, notice the number of closed depressions on the flow, some elongate, others irregular, most about ten feet deep. Rubble and sand cover their floors. They indent a relatively flat lava surface and flows tilt down around their edges, so they appear to be collapse features. They probably formed as molten lava within the flow drained out from beneath a solidifying crust.

Continue driving along the haul road, and stop where it curves west one mile from the starting point. Here the road crosses a small lobe of lava that flowed down a shallow gully to the west. Basalt lava is so fluid that it follows minor topographic features. A basalt lava flow can be diverted with an artificial embankment, as has been done in Hawaii and Iceland with varying degrees of success. Walk east onto the flows. At first glance, they look younger than those at the last two stops because it appears that less sand covers the surface. But the surface is covered with platy chips of basalt, which also cover patches of sand that fill the low places. Sand is actually more abundant than appears at first glance. These lavas are probably about the same age as those at the preceding stop.

Drive on to a large parking area just outside the gate at 1.5 miles from National Trails Highway. Just inside the gate, and 100 feet to the east, is

Collapse depression in Pisgah lava flows, floored by windblown silt, off Lavic Road, 3 miles south of National Trails Highway. Sunshine Crater (another young cinder cone) and Sunshine Peak in background.

a dark, rough, young-looking lava flow with very little sand cover. It has a nice aa surface. Other flows in the vicinity have ropy pahoehoe surfaces.

Return to the paved highway. If you would like to see the differences between lavas produced during the first and second periods of Pisgah's eruption, drive two miles east to where the highway bends north (left) to cross the railroad tracks, and park. Walk south onto sand-covered lavas. If you chip out a piece of lava and examine a fresh surface, you will notice that few crystals are visible without a hand lens. This is characteristic of lavas from the first eruptive period. These lavas are overlain by an extensive flow from the second period. Walk a little farther, and climb up onto the lava bench. Look at a piece of lava; it has abundant crystals, about ¹⁄₁₆" to ⅛" across, of white plagioclase feldspar and clear green olivine. Rocks from the third eruptive period have even larger and more abundant crystals.

Two other dirt roads give good access to the lava flows. One is the power line road exactly one mile west of the haul road; do not confuse it with another power line road only 0.1 mile west. This passable road trends southwest along a triple set of mammoth power lines that carry electricity from Hoover Dam to Los Angeles. The road forks 0.2 mile from the National Trails Highway; bear left and follow the lines. Lava flows first appear in about two miles. The road crosses a wide variety of lavas, including some with excellent ropy surfaces. Large piles of pale dirt a mile to the west mark an open pit mine that produces hectorite, a rare clay that contains lithium. It comes from playa lake deposits beneath the lava flows.

The other access, Lavic Road, heads south at a right angle from the pavement 4.8 miles east of the haul road, where National Trails Highway turns 90 degrees north (left) to cross Interstate 40. This rough and rocky road is passable for passenger cars, and the drive is worth it. Bear right at the T-intersection just across the railroad tracks. Follow Lavic Road approximately 3 miles and park near the edge of prominent lava flows. Don't drive into the Marine Corps base. Hike a short distance west to the lava flows, and scramble onto them. The flows form a broad plateau pocked by numerous depressions, some as much as 50 feet across. These are collapse sinks; they formed as molten lava drained from beneath a hard crust.

Tilted lava flow at margin of collapse depression, off Lavic Road, 3 miles south of National Trails Highway.

Young basaltic volcanoes, such as Pisgah and Amboy, are common in the desert regions of western North America. Why they occur there (and not, say, in Nebraska) is not well known. Why they pop up exactly where they do in the deserts is also unclear; few can be tied directly to a fault or other feature that might localize the eruption. This situation is quite different from subduction-related volcanoes, such as the Cascades, that occupy specific, predictable positions relative to the continental margin.

At present, no one can predict when and where the next eruptions might occur. An eruption in an unpopulated area would be a spectacular fireworks show; an eruption in a town or through a hazardous waste dump would be a disaster. Imagine the scene if a volcano like Pisgah were to erupt in Las Vegas or Lake Mead! For these reasons, the young volcanoes of western North America will remain under close scrutiny.

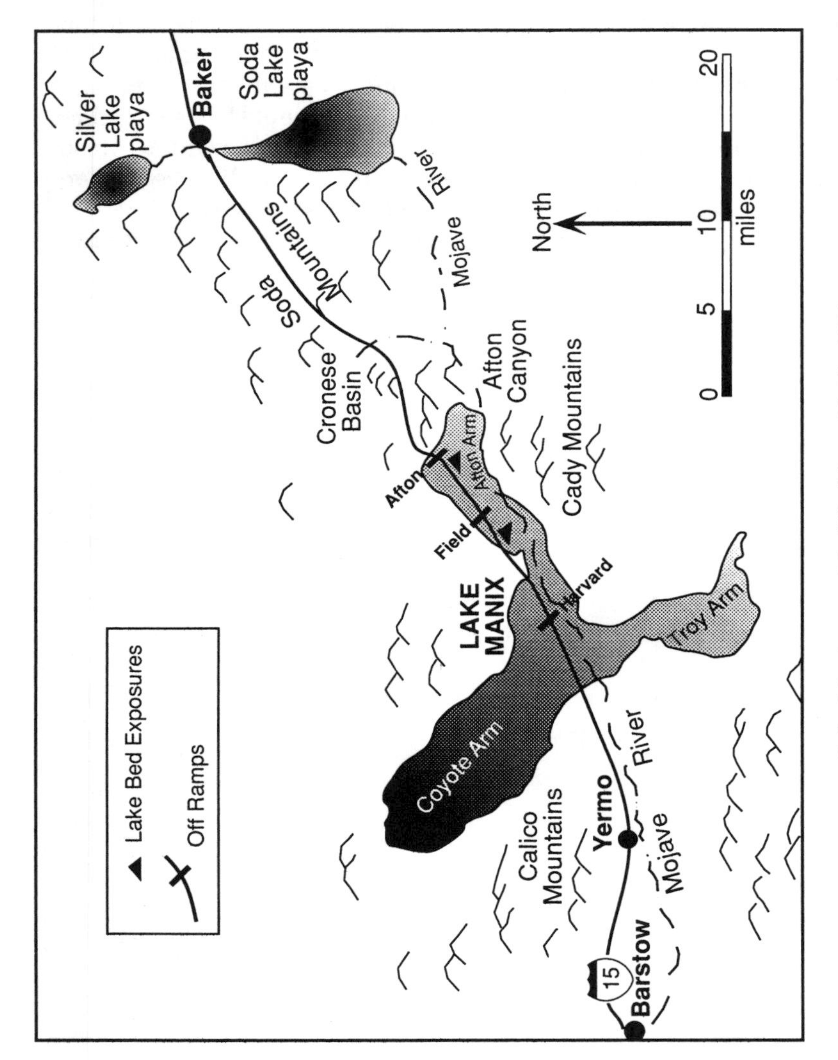

Lake Manix area. Modified from Meek, 1989.

VIGNETTE 15

FLAMINGOS IN THE DESERT

Pluvial Lake Manix

SAN BERNARDINO COUNTY

The Mojave Desert was not always as desolate and dry. About 15,000 years ago, not long ago geologically speaking, streams flowed through the arid and semi-arid terrain nourishing blue-water lakes in the broad valleys. The cooler and wetter climates of the ice ages were pluvial periods of increased rain in southern California. Pluvial water from the Sierra Nevada ran all the way to Death Valley through the Owens River and a string of four large lakes. The Mojave River, flowing out of the San Bernardino Mountains, first north, then east, and finally north again, also reached Death Valley through a succession of at least three large lakes. Death Valley, its floor 280 feet below sea level, was the ultimate sink for much of our southwestern desert, including the lofty (12,000 feet) Spring Mountains of western Nevada, which still feed water to Death Valley through the Amargosa River. Not surprisingly, Death Valley harbored a pluvial lake more than 100 miles long and 600 feet deep.

Those desert lakes must have been oases. Those with through-flowing streams were fresh, but lakes with no outlets gradually turned brackish and eventually saline as evaporating water left its dissolved salts behind. The pluvial lakes dried up with the change in climate after the last ice age ended some 10,000 years ago, and they all became brackish or saline in their final stages. Those desert lakes must have been beautiful in their prime; their placidness and deep blue color contrasted with the harshness of the surrounding rough and relatively barren terrain. Their shorelines, lush with bulrushes and other plants, provided rich habitats for birds, mammals, reptiles, amphibians, fish, insects, and shelled creatures. Native Americans lived along the lakeshores. Mono Lake, east of the Sierra Nevada, and Walker Lake, in west-central Nevada, are two modern desert lakes that give some sense of what the vanished pluvial lakes were like.

Much of the water for desert lakes came from melting snow and ice in nearby high mountains, as well as from local rain and runoff. Such water fed the pluvial Mojave River, which nourished Lake Manix. Fifteen thousand years ago, the San Bernardino Mountains received heavy snows that created a handful of small glaciers. Spring and summer melting sustained a perennial Mojave River all the way to Death Valley. Today, the Mojave River is an ephemeral stream that fills much of its river bed when the snow melts. Even so, people who live along it must reckon with floods when an unusually warm spring follows a winter with heavy snow in the mountains. Such floods carry water all the way to Silver Lake playa north of Baker, flooding that community en route. Several feet of water can accumulate in Silver Lake and linger for the better part of a year. The historical record is ten feet during the winter of 1916, an abnormally wet year. No historical records tell of the Mojave River having flowed all the way to Death Valley, but the path it followed north from Silver Lake during pluvial times is clear. Water in Silver Lake must be 36 feet deep to overflow the sill at its north end and spill into the old channel the Mojave River once followed to Death Valley.

The Mojave River is deceptive. Even in the driest time of year, when no rain has fallen for months and the surface of its bed is bone dry from the mountains to Victorville, water flows through the narrows at Victorville. California 18 crosses it there, as does Interstate 15 a little farther north. The coarse sand in the river's bed is so pervious that water percolates into and through it easily. The river goes underground in dry periods, and its water moves slowly downstream through the porous sand in its bed. This is hard on fish, but it conserves water by slowing runoff and reducing loss from evaporation. At Victorville narrows, a nearly impermeable mass of granite forces the water to surface. Within a mile or two downstream, it soaks back into the streambed, except during floods when surface flow extends far into the desert.

Beyond Yermo, about 15 miles east of Barstow, both Interstate 15 and the Mojave River, which follow roughly parallel courses north of Victorville, enter the broad valley that pluvial Lake Manix once flooded. The highway crosses the abandoned lake bed eastward for 19 miles, to the Afton exit.

At its maximum filling, Lake Manix covered about 85 square miles and had three large bays. Coyote Arm was north of the present location of Interstate 15, tucked in behind the Calico Mountains. Troy Arm was farther south, along where the Santa Fe Railroad and Interstate 40 are today. Afton Arm was where Interstate 15 is today. People rushing to the gaming tables of Las Vegas travel through the heart of the Lake Manix basin and Afton Arm.

What evidence tells of a lake that once occupied this basin? The extremely flat valley floor is suggestive, but many desert basins, never occupied by lakes, have flat floors. Fortunately, the Mojave River has cut 100 to 200 feet into the floor of Afton Arm exposing the underlying deposits. You can see that the uppermost beds are mostly fine sand, silt, and clay—typical lake deposits. They rest, locally unconformably, on older lake beds and stream-laid sand and gravel, deposits that are not part of the Lake Manix sequence, although they accumulated in the same basin. These pre-Lake Manix beds contain at least three ash layers, of which only the lowermost—the 2.2 million-year-old Huckleberry Ridge ash from Yellowstone—is correlated and dated with confidence.

The one identifiable ash layer within the Lake Manix beds is correlated with a 185,000-year-old tuff in the southern Sierra Nevada. Its position near the bottom of the sequence indicates Lake Manix came into existence about 190,000 years ago.

The riverbank exposures are not visible or easily accessible from Interstate 15. But gullies draining to the Mojave River did erode through the lake beds in two places along the highway. Eastbound travelers get the better view. One exposure is just south of the highway and north of the parallel railroad tracks 5.2 miles east of the Harvard Road exit, where the

Oblique air view south of large beach ridge (center) along east end of Afton Arm of Lake Manix, near Afton exit on Interstate 15. —From *Geology Illustrated*, by John S. Shelton, W. H. Freeman and Company, copyright © 1966.

Platy beach pebbles in a small excavation near the south end of Afton beach off Interstate 15. Knife is 5½ inches long.

highway dips gently into a broad swale. The lake beds are soft, fine-grained, and slightly greenish. Westbound travelers should watch for this site 2.2 miles west of the Field Road exit, as they start down into the swale. A larger exposure of lake beds, clearly visible to motorists going in either direction, is 3.5 miles east of the rest stop and 0.8 mile west of the Afton Road exit. You can reach it on the frontage road west from the Afton exit.

Other souvenirs of pluvial Lake Manix include lake shoreline features such as wave-cut cliffs, terraces, and beaches. To see the very best beach, take the Afton exit, and then drive a hundred yards or so south onto a wide and level ridge. The ground along the ridge's western edge and the face of the ridge are littered with smoothly worn, flat pebbles, some nicely circular. These are typical beach pebbles. They would make good skipping stones, if we had some water. The ridge is a magnificent beach built by large waves driven east up Afton Arm by the powerful westerly winds. This beach ridge is part of the highest young shoreline of Lake Manix. Its size indicates that the water stayed at that level for a long time, presumably because the outlet stream flowed over hard rock that it could not erode. That high outlet may have been through Troy Arm. About 15,000 years ago, an outlet developed through Afton Arm, a little east of here. Water pouring across soft rock quickly carved Afton Canyon, putting the presumed Troy Arm outlet out of business and sending water through Soda and Silver Lake playas and Silurian Dry Lake to Death Valley.

We have discussed only the latest events in the history of Lake Manix. If you hiked into gullies off the frontage road, you probably noticed layers of brownish sand and gravel within the deposits of lake clay. They are stream deposits, which show that Lake Manix shrank dramatically at times, perhaps disappeared. Age dates on the ash layers within the older lake beds, indicate that water flooded the basin at least 185,000 years ago,

Lake clays in gully bank just south of frontage road, west of Afton exit from Interstate 15. Pocketknife is 5½ inches long.

long before the beach ridge at Afton exit formed. The record in the landscape tells only of the last chapter in a long and complicated history.

The fossil remains of animals and birds that lived along the shores of Lake Manix are preserved within shoreline deposits. Animals represented include dogs, cats, bears, horses, camels, antelope, bison, sheep, and mammoths. Most were grazing animals. Shoreline birds were abundant, including storks, pelicans, cormorants, grebes, ducks, geese, eagles, cranes, and even two species of flamingos. Imagine them standing on one slim leg, solemnly surveying the Lake Manix scene. It really happened. Under the flamingos' feet were snails, beetles, fish, and an occasional turtle. This was a thriving community.

The Mojave Desert must have been a rather inviting place during pluvial times, 15,000 and more years ago. It was a land of lakes, streams, and wildlife, a land that locally supported grass, flowers, bushes, and even small trees such as piñon pines and junipers. Its size and the strong westerly winds would have made Lake Manix a great place for windsurfing—had any of the local denizens been so inclined.

VIGNETTE 16

EIGHTY SECONDS OF CATASTROPHE
The Blackhawk Slide

SAN BERNARDINO COUNTY

Anyone viewing the north side of the San Bernardino Mountains south of the western Lucerne Valley will be impressed with their high, abrupt face, and the broad alluvial apron that slopes smoothly northward from their base at an angle of two to three degrees. The face is approximately 3,000 feet high, and the transition to the alluvial apron is sharp along a nearly straight line. The steepness of the mountain face, the sharp line of its base, and its straight trend all suggest that it is the line of a fault that raised the mountain block above the lower desert terrain. Vignette 13 contains a discussion of stresses and structures in this area.

Ten miles east of the Lucerne Valley settlement, a lobate mass of shattered rock extends from the mountains far across the alluvial apron. This is the famous Blackhawk slide, named for the high mountain on this part of the face of the range. The lobe is a run-out slide, a thin sheet of debris that moved a long distance across a gentle slope.

California 247 passes within 500 feet of the end of the slide. Its edge, 50 feet high, can be approached on a short, graveled road to quarries in the terminal part of the slide. This road goes south where a wooden-pole power line from the east turns north along Santa Fe Road. The lobe of slide debris extends as a coherent sheet an incredible four and a half miles out from the mountains on a surface sloping only two to three degrees northward. The lobe's average width is one and a half miles, the maximum is two miles, and thickness is 30 to 50 feet. Locally, mostly to the west, the face of the slide has a discontinuous mantle of recent windblown sand and silt; don't be misled by it.

You can get there by taking California 18 and 247 east for 31 miles from Victorville, or by taking California 247 north and west for 36 miles from Yucca Valley, or by traveling south and east 42 miles on California 247 from Barstow. People coming from the south can shorten their journey by taking Bear Valley Road cutoff east from Interstate 15 to a junction with California 18 east of Apple Valley. Travelers approaching from the west by way of California 18 continue straight ahead onto Highway 247 in the settlement of Lucerne Valley and follow it east 10 miles. Refer to a road map of San Bernardino County, the U.S. Geological Survey, 7.5-minute, Cougar Buttes topographic quadrangle, and the San Bernardino sheet of the California Geological Map.

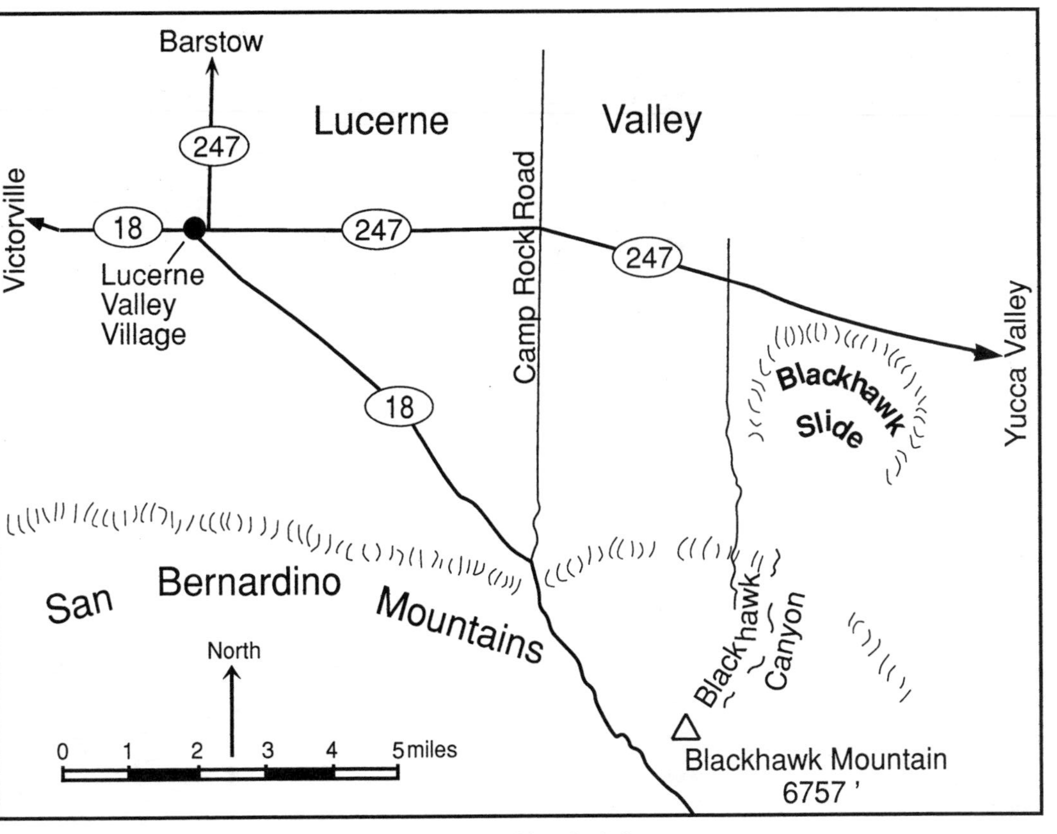

Access to Blackhawk slide.

Oblique air photo, looking south, of Blackhawk slide lobe. Note the abrupt gullied margin, elevated terminal ridge, transverse wrinkles, alluvial mantle on upper part, and prominent east-side linear ridge. —John S. Shelton photo (taken in 1959)

People interested in the slide's surface can ascend on foot from the west quarry or any of the small gullies indenting the lobe's margin. What they see is a hummocky surface of rounded hillocks and ridges, with interspersed longitudinal furrows. Scattered, undrained depressions are floored with fine, light-colored pond deposits, laid down after sliding ceased. A rimlike ridge at the terminal edge rises several feet above the slide surface but is more obvious on air photos than in the field. Likewise for the rounded ridges that extend transversely across the slide, giving its lower half a rumpled appearance on air photos. Walking on the slide's surface is easy, allowing you to roam widely.

Much of the upper three miles of the lobe is mantled with alluvium derived from the adjacent mountains. Local islands of slide surface projecting through this mantle suggest that ridging may continue almost to the head of the lobe, although the ridges are fewer and more longitudinal than transverse, in contrast to the lower reaches of the slide's surface. Prominent are two ridges forming lateral borders for the upper half of the slide. The east-side ridge attracts particular attention for its linearity, but the west marginal ridge is as high, 50 feet above the slide surface, although more irregular.

East of the lower part of Blackhawk slide are granitic hills rising several hundred feet above the alluvial apron. One, identified as 3735 from its elevation (3747 on old maps), was overridden by the slide, which ascended at least 250 feet. This hill is easily identified by its smoothly rounded summit and the many gullies riving its slopes. It lies to the south-southeast of the quarries.

Gullies dissecting the terminal margin of the lobe and quarry exposures show that it consists of angular fragments of shattered rock, which geologists call breccia. Exposures are best in the walls of the west quarry, but do not get too close to vertical quarry faces. They are unstable and can collapse at any time. If the quarry is being worked, you may not be admitted, but if it is inactive, proceed with care. One sees that the breccia

Vertical air photo of Blackhawk slide lobe. Compare with map of features for identification on page 151. —Pacific Aerial surveys (taken in 1988)

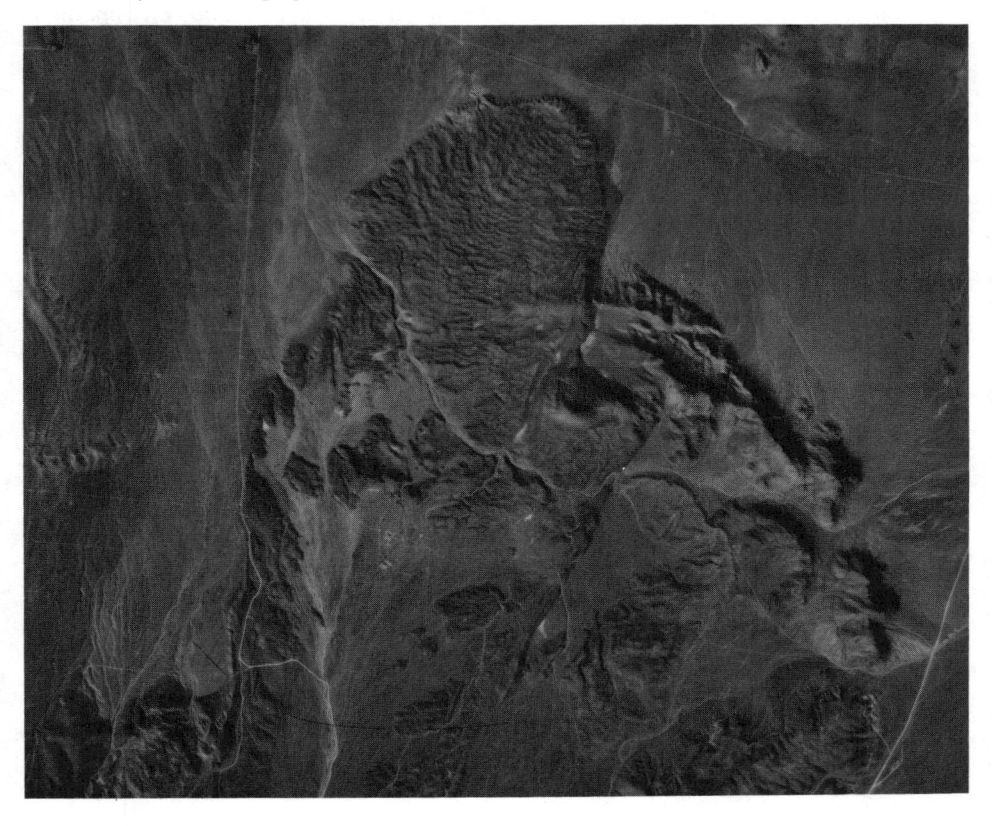

consists almost entirely of fragments of a gray, calcareous rock called marble. It is recrystallized limestone and composes much of Blackhawk Mountain, where it is known as the Furnace limestone or marble. The average fragment size in the breccia is about one inch, but angular cobbles of three to six inches are abundant, and the prevailing range is from a powder to about ten inches. Scattered larger stones of three to ten feet exist, the largest reported being 35x25x15 feet—not a trivial fragment. The color banding and layering of what were once large, unbroken

Features of Blackhawk slide.

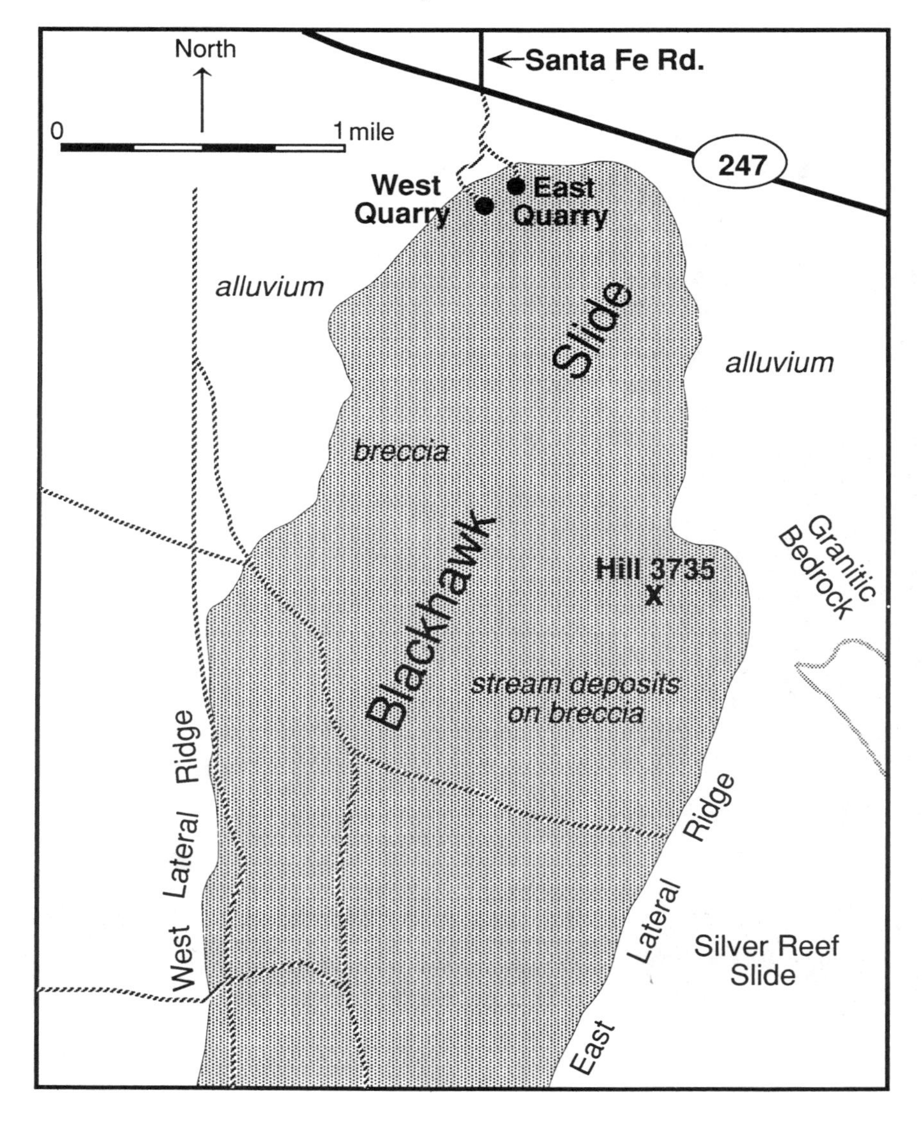

Marble breccia exposed in wall of west quarry. Hiking staff is 4 feet long.

Folds within slide breccia near the terminus of the slide are exposed in a wall of the west quarry. Four-foot staff near center of down-fold provides scale.

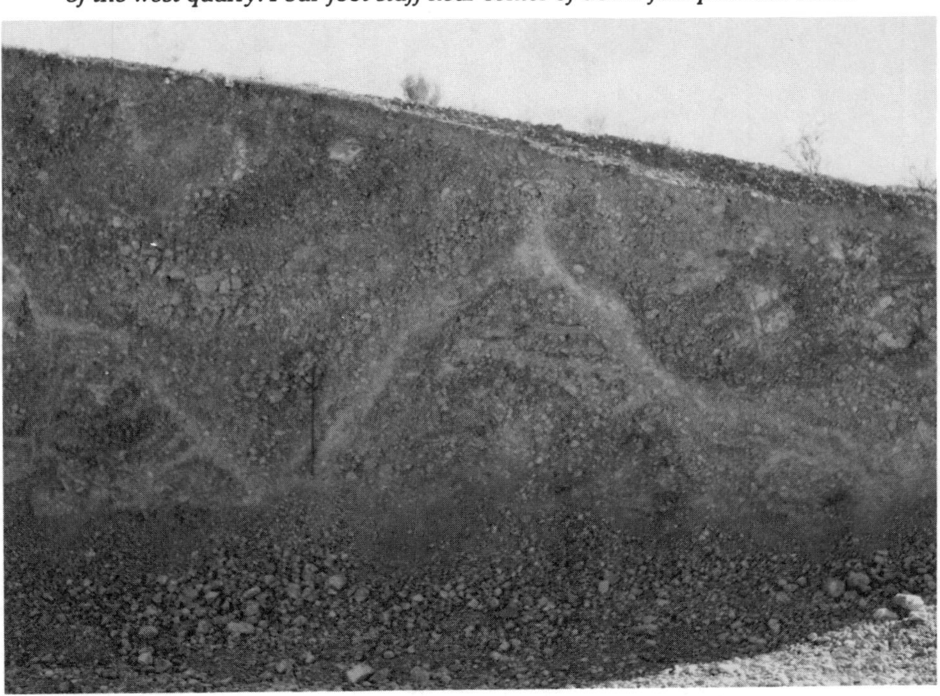

chunks of marble can be traced through closely nestled small fragments for many feet, with little disturbance aside from tiny cracks between fragments. Such assemblages are called domino breccia. We will consider their origin and significance a bit later.

Although monolithologic (one type of rock) marble breccias are dominant, breccias of other rocks are present and aid our understanding of the slide. At the base of the lobe's terminal and lateral edges lie scattered masses of brownish sandstone, one of which is well exposed in

This cut along the access road to the east quarry reveals Old Woman sandstone with a lens of marble breccia at the terminus of the slide. Four-foot staff at lower left.

a west-side cut along the road into the east quarry. Lenses of similar material crop out in some gully walls within the slide. This sandstone is derived from the Old Woman sandstone, a rock formation underlying Furnace marble in the north face of Blackhawk Mountain. In a few spots within the slide, underneath the sandstone, small masses of brecciated metamorphic rocks (gneiss and quartzite) and coarse-grained granitic rock are exposed. Those rocks underlie the Old Woman sandstone in the mountains and were presumably carried down with the sandstone during the catastrophic slope failure that produced the slide.

The slide lobe contains about ten billion cubic feet of crushed rock. If you have ever spread just one cubic foot of crushed gravel, you have some feel for the colossal task accomplished by the Blackhawk slide.

We can draw several conclusions from the composition and structure of the slide sheet. First, the stratigraphy of rock formations exposed in the source area, specifically Furnace marble on top of Old Woman sandstone, resting on weathered and disintegrated gneiss, quartzite, and granitic rock, is preserved within the slide. The units are much disturbed and attenuated, but they are in proper sequence. Second, the bulk of the slide moved as a sheet without significant internal mixing, as would have occurred had it flowed. Its ascent of 250 feet to overtop Hill 3735 required a velocity calculated at not less than 75 miles per hour and possibly as much as 170. We can further calculate that if the slide had a launch speed of 170 miles per hour near the head of the alluvial apron, as suggested by its fall down the face of Blackhawk Mountain, it could have accelerated to a maximum velocity of 270 miles per hour descending the alluvial apron. The entire journey from launch to terminus would require only 80 seconds at these velocities. The run-out of four and a half miles on a slope as gentle as two to three degrees indicates minimal drag, superb lubrication, or some exceptional mode of movement at the base of the slide.

The domino breccias, the abrupt margins, the surface wrinkles, and compressional folds within breccias exposed in the west quarry all suggest that the slide stopped abruptly. Domino breccia represents a large block of rock carried as a coherent piece until the moment of impact, which shattered it into thousands of fragments. These fragments could not possibly have been carried as separate pieces from the source to their present site and still maintain their original relative positions. Transverse ridges on the slide's surface, including the terminal rim, are attributed to compression and thrusting up as the terminal part stopped while the remaining body continued to push forward. Cessation of movement progressed backwards like a wave moving upstream.

The slide sheet was dry, both internally and at its base. No indication of water run-out from the margins or any suggestion of mud at the base or within the slide have been seen. The slide may have generated a blast of air beyond its terminus, as reported for similar events of historical date, but we find no concrete evidence to support this. The sharpness of the slide's edge is impressive. Individual rocks were not ejected onto the surrounding terrain, and except for enveloping dust, and possibly a severe air blast, persons standing a short distance from the terminus might have survived unscathed, although probably newly gray-haired.

So, where did this huge sheet of shattered rock come from? How did it move so far over a two to three degree slope? When did this happen? What was the cause? And could it happen again?

The source is obviously the steep, rugged front of San Bernardino Mountains to the south. The Furnace marble of Blackhawk Mountain matches perfectly the marble fragments of the breccia, and the brownish sandstone, gneiss, quartzite, and granitic rock, like the fragments at the base of the slide breccia, are also exposed in Blackhawk Mountain. Owing to a thrust fault, one of many that cut the range, the tough, dense Furnace marble rests on top of the younger, weaker, poorly consolidated Old Woman sandstone—a common thrust fault relationship but opposite to the normal depositional relationship.

View south-southeast across the surface of the Blackhawk slide to Hill 3735, with rilled slopes. The slide overtopped this hill. The light streak in midground is the dry surface of a playa pond.

To have a heavy, massive rock overlying soft, friable strata, and both exposed in the face of a steep mountain slope, is a prescription for disaster. The weak sandstone weathers rapidly to a crumbling mass, undermining (sapping) the overlying tough, cliff-forming marble. This undercutting, probably aided by groundwater seepage, continually steepens the slope until it collapses. The slide may have been triggered by an earthquake, of which southern California has a surplus, but the stage was set by these more mundane erosional processes.

From the head of the slide scar on Blackhawk Mountain, the rock debris descended 2,000 feet down a steep slope before taking off as a sheet across the alluvial apron, which it descended another 1,500 feet. The total fall was 3,500 feet, and the total distance traveled was 5.7 miles. Movement of the debris sheet for four and a half miles across a very gentle slope suggests that its base was unusually well lubricated, or that it was endowed with a highly facile mode of flow or slip within a thin basal layer.

Steepness of the Blackhawk Mountain face, and comparisons with historically observed slides, suggest that the rock debris initially moved as a rockfall. At the mountain's foot, it crossed a ridge of gneiss with a steep face on its downstream side. According to one model, this configuration launched the descending sheet into the air several hundred feet above the ground. As it settled back, the debris mass captured some of the underlying air and compressed it into a basal layer about one foot thick. The lubricating medium for the slide sheet crossing the alluvial apron was possibly this thin layer of compressed air. If trapping of air under a slide seems unlikely, remember that the slide was moving very fast, and the air needed to be contained for only about 80 seconds. Fine-grained Old Woman sandstone, weathered gneiss and granitic rock, as well as some surface soils within the basal slide debris may have formed a membrane that helped impede the escape of air. Heavy suspension of dust within the basal air layer might have had a similar effect. Early grounding of lateral margins of the slide lobe, as suggested by the height of the marginal ridges, would have helped confine the air layer. These marginal ridges are high relative to the surface of the slide because they were formed early, when the leading, thick part of the debris sheet was here. It moved on, leaving the grounded marginal ridges standing higher than the following thinner part of the slide sheet. Slide velocity at the launch point is calculated at 140 to 170 miles per hour, and at that speed not much air escaped from the leading edge of the sheet.

The concept of lubrication by a thin basal layer of compressed air is attractive. Still, not everyone likes the idea. Another suggestion is that high-frequency sound waves, generated by a sliding mass, could keep the particles of a fine-grained basal layer separated, thus creating a fluid-like condition that permitted easy and rapid movement. The slide sheet simply went along for the ride on the back of a thin, fluidized, basal layer of fine particles. This concept merits consideration too, although it does not explain some features of the Blackhawk slides as well as the air lubrication mechanism. The acoustical lubrication model works better for large, presumably high-velocity slides on Moon's surface, where there is no atmosphere, and it may also apply to the huge slides of Mars, where the atmosphere is less than one percent as dense as on the earth.

How old is Blackhawk slide? Thanks to a radiocarbon (carbon-14) age obtained from freshwater snail shells from pond deposits in depressions on the slide, and the radiocarbon age of a post-slide rock varnish on surface boulders, we know that the slide cannot be younger than about 17,000 years. The degree of gully dissection of the terminal margin of the slide, and the amount of alluvial debris mantling the upper two-thirds of

Surface of playa pond in a closed depression on the slide. Shells for radiocarbon (carbon-14) date of the slide came from such a pond.

its surface are indicative of a similar time interval. It is a reasonable estimate that the Blackhawk slide occurred 17,000 to 20,000 years ago.

Have similar slides occurred here before? Indeed they have. Along the east margin of Blackhawk slide are remnants of breccias laid down by an older slide of smaller dimensions, three to four miles long by two miles wide, and with a slightly different direction of movement. It is the Silver Reef slide. Furthermore, remnants of marble breccia low on the flank of Blackhawk Mountain, which are now buried beneath alluvium known to be older than the Silver Reef slide, provide evidence of at least one still older slide. It has no recognized topographic expression in the current landscape.

Could another Blackhawk-type slide happen again in the future? Certainly, inasmuch as the basic conditions favorable to sliding and the processes creating over-steepened slopes are still at work. The geological record suggests, however, that such a slide is not likely to occur soon—so don't let that consideration deter you from visiting this extraordinary site.

Blackhawk slide is one of the most easily observed and spectacular examples of a long run-out slide extant. It is not unique, but it is a beautiful sample of the species. The slide lies within an easy one-day trip from densely populated southern California areas. Pack a picnic lunch and make a day out of a loop trip via Cajon Pass (Vignette 11), Victorville, and Apple and Lucerne valleys to the slide. Then proceed east and south via California 247 to a junction with California 62 a little east of Yucca Valley. Follow California 62 west and south to Interstate 10 in Coachella Valley, and proceed west through San Gorgonio Pass between the two, often snow-capped sentinel peaks, San Jacinto and San Gorgonio, to your home area. All roads are paved, you circle San Bernardino Mountains, see some fine, high desert country, and become acquainted with an unusual geological feature.

Recognition of the landslide origin of the Blackhawk breccia lobe was first published in 1928 by Pomona College's revered geologist, the late A. O. Woodford, and his associate, T. F. Harriss. The most thorough study of the slide and the suggestion that it was air-lubricated comes from the Ph.D. research of Ronald L. Shreve, now a noted U.C.L.A. Professor of Geology and Geophysics. His study stimulated H. Jay Melosh, now of the University of Arizona, to suggest the acoustical fluidization mechanism for similar slides on Earth, Moon, and Mars. Many geologists have poked around, admired, and commented on the Blackhawk slide, but none has made contributions commensurate with those noted above. Ron Shreve is the guru of the Blackhawk slide.

VIGNETTE 17

A HUGE PILE OF LIVING SAND

Kelso Dunes

SAN BERNARDINO COUNTY

Exploring dunes is fun. They can be mysterious and beautiful. Many are alive, changing shape with each shift in the wind. The witching time for dunes is just before sunset and just after sunrise, when shadows are long and deep. Then one appreciates the remarkable grace of curving dune crests and the complicated intermingling of geometrical forms. Walking through dunes at sunset, when the western sky reflects its twilight glow onto the sand, can be a mystical, magical, almost religious experience.

Kelso Dunes lie in the eastern Mojave Desert, about midway between the diverging paths of Interstate 15 to the north and Interstate 40 to the south. Kelso is an abandoned section station of unusual elegance, size, and history on the Union Pacific Railway. It was established in 1906 as a station along the San Pedro, Los Angeles, and Salt Lake Railroad, which later became a subsidiary of Union Pacific. Best seasons to visit the dunes are winter, spring, and fall.

Kelso Dunes rise from a broad alluvial apron that slopes very gently down to the north from the Granite Mountains. Projection of this smooth surface under the highest dunes suggests a sand thickness of at least 700 feet. The dunes lie at the end of an umbilical cord supply line of wind-blown sand four to five miles wide that extends 35 miles east and southeast from the mouth of Afton Canyon (see Vignette 15). This long trail of sand includes the sand-covered Devil's Playground just northwest of the dunes, which receives additional sand from the north. The rugged

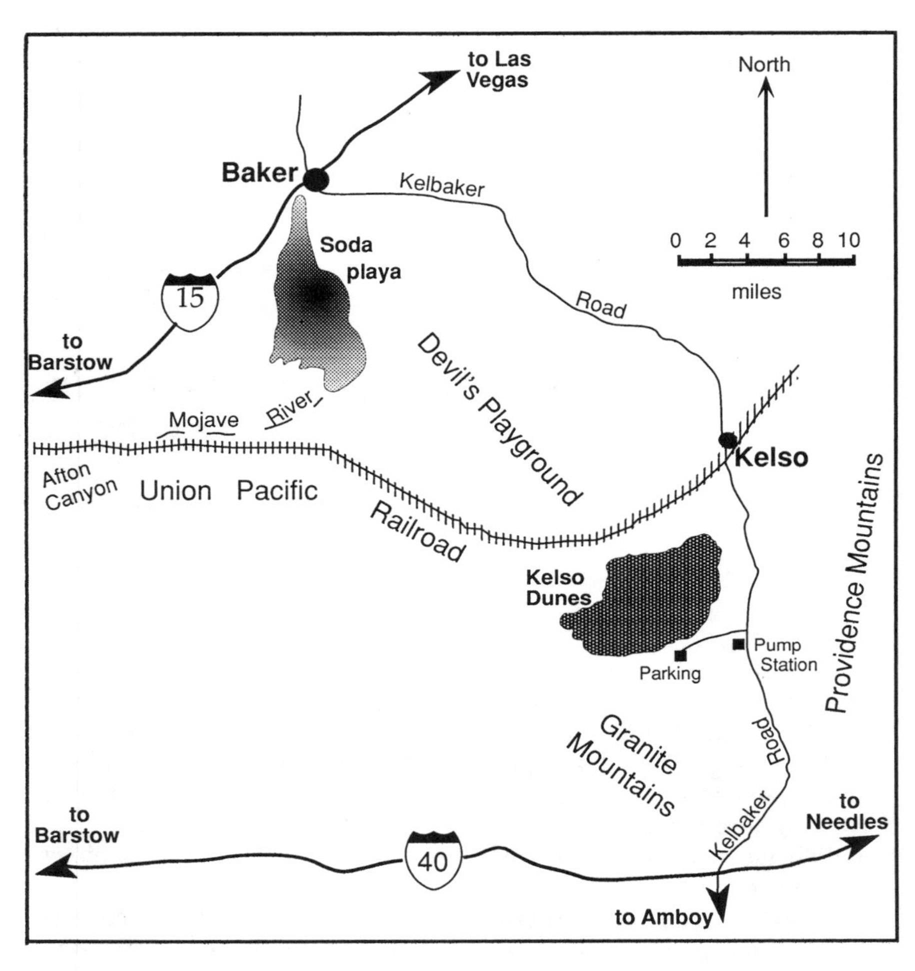

Location of and access to Kelso Dunes.

Roads are shown on auto club maps of San Bernardino County. The eastern part of the dunes appears on the U.S. Geological Survey, 7.5-minute, Kelso Dunes topographic quadrangle. To get there from the south, take Kelbaker Road north about 12 miles from its intersection with Interstate 40. Access from the north is longer, 44 miles, from Baker on the Kelbaker Road. That route skirts the geologically young Cima volcanic field, with 40 cinder cones and many lava flows.

Turn west from Kelbaker Road onto a well-graded dirt road marked by a Bureau of Land Management sign a little north of a gas pipeline pump station. Drive west about 3.5 miles to an obvious parking area close to the dunes and nearly due south of their highest point. Walk from here. Dune buggies and all-terrain vehicles are prohibited, and your car would get stuck.

Kelso Dunes, as viewed from the parking area looking north. The highest point, on left-center skyline, is 500 feet above camera.

Providence Mountains to the east were named by early Mormon travelers in appreciation for their springs.

Prevailing winds in the Mojave Desert blow from the western quadrants. Here at Kelso Dunes, the shape of the landscape channels the winds into the prevailing west-northwesterlies, although ripple marks and dune forms show that strong winds occasionally attack the dunes from all directions. These storm winds cause profound changes in dune shapes.

Four large, linear ridges that bear east to northeast are the main features of the Kelso dunes complex. Many smaller transverse ridges cover their flanks and the areas between them. A typical transverse ridge is 200 to 300 feet long. Its broad windward flank slopes 10 to 15 degrees and has a firm surface commonly decorated with little ripples. The windward slope climbs to a smooth crest, which drops off into a steep leeward face that stands at the angle of repose of dry sand, up to 34 degrees. The upper edge of the steep lee face generally ends a bit lower than the dune crest, which typically is gently rounded. Lee faces range in height from a few to 30 feet and vary along a dune's length. The somewhat sinuous crestline alternates between broad, rounded summits and open intervening saddles.

The wind rolls some sand grains along and bounces others across the dune surface in a series of short hops. The bouncing grains knock others across the surface as they crash-land back on the dune. Those impacts

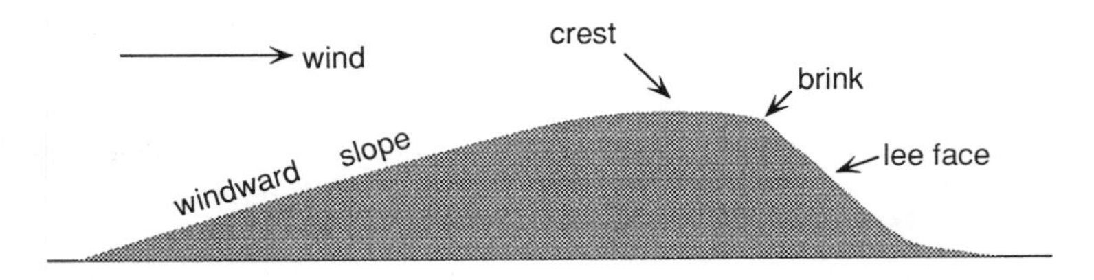

Cross section of a typical transverse dune ridge.

kick large grains along in a series of sudden jerks, like a small boy kicking a tin can down the sidewalk. Bouncing, the most efficient means of wind movement of sand, accounts for about three-quarters of the total sand transport. The difference between rolling and bouncing resembles the difference in mobility between a child that crawls and one that walks. The latter covers more ground and gets into twice as much mischief.

Bouncing sand grains cover distances up to several feet with each hop, angling back into the dune at nearly the speed of the wind. When a strong wind blows across a dune, you can see the cloud of bouncing sand grains blurring the view of its crest and windward slope. The cloud of sand is rarely more than 12 to 16 inches high, so you can stand comfortably in it, and someone with bare legs can tell you how high the grains are bouncing. Avoid the lee side of the dune, where you will be deluged by a rain of falling sand grains.

Sand grains blowing up the windward slope of a dune enjoy a free ride at the expense of the wind until they reach the brink where the lee face drops off. Here, grains moving along the surface are unceremoniously dumped down the lee face, and the bouncing grains make their last leap, coming to rest somewhere below, sheltered from the wind. The net effect

Modes of wind-driven, or eolian, transport.

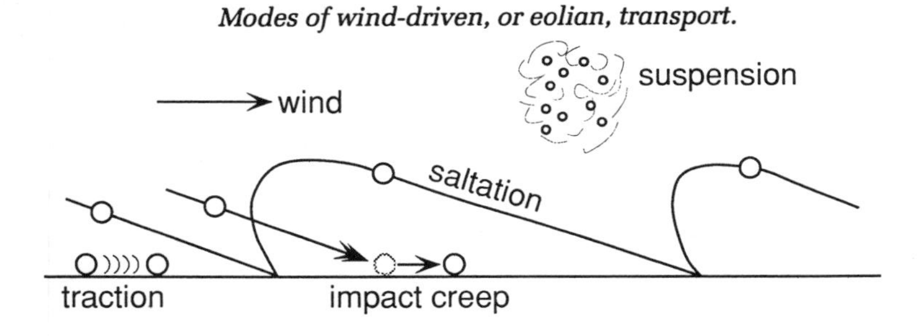

of removing sand from the windward slope and depositing it on the lee slope is to cause the dune to migrate, or advance, downwind.

Since the cloud of bouncing grains is denser near the bottom, you can be sure that most grains are bouncing to low heights and consequently making short leaps. As a result, most of the bouncing grains that cross the crest land on the upper part of the lee face, making it steeper. When it reaches an angle of about 34 degrees, the upper part of the lee face slides in a tongue of avalanching sand toward the bottom of the slope. When strong winds are moving sand, those avalanches occur repeatedly on the lee face of a rapidly advancing dune. If you visit dunes shortly after a storm wind, you will find it easy to generate avalanches by stomping along the upper part of the lee face. It is fascinating to watch them run down a lee slope. Nearly all sand on lee slopes has been moved by avalanches.

A transverse dune, with its crest at right angles to the wind, is an efficient sand trap; very little sand escapes it. Wind that has passed over a dune, having lost much of its sand, starts to pick up sand from the succeeding hollow and from the lower windward slope of the next

Lee face of a transverse dune after heavy wind, showing fresh and partly healed avalanche tongues, all of natural origin.

downwind dune. By the time it gets part way up the windward slope of the succeeding dune, the wind is loaded with sand and ready to deposit part of its burden. This type of deposit is called accretion sand. Experienced dune hikers know that accretion sand is made firm by the impact of bouncing grains. You hardly leave a footprint in it. The upper part of the windward slope has the thickest and firmest mantle of accretion sand. However, near the brink of a rapidly advancing dune, accretion sand may be so thin that a hiker breaks through into the soft, avalanche sand on the lee slope beneath. Avalanche sand is so loosely packed that a walker sinks in up to the ankles. It is smart to keep on, or a little windward of, the dune crest for easiest walking.

Like those of dunes everywhere, sand grains in the Kelso Dunes are smoothly worn and nicely rounded. If you look at the grains under a

Bedding in wind-scoured, frozen Kelso Dunes sand.

Wind ripples on sands of different grain size: background 2 mm and larger, foreground 0.3 mm.

strong hand lens or microscope, however, you will find that the almost perfectly round surfaces are covered with tiny pits caused by the impact of one grain against another. Grab a handful anywhere, and you will see that the grains are all about the same size, a result of the powerful winnowing force of the wind. Most of the sand is composed of the minerals quartz and feldspar. Thin layers of black sand, mostly the magnetic iron oxide mineral magnetite, catch your eye, especially where they make a patch on the surface rather than just a streak. Children enjoy dragging a magnet through the black sand, covering it with magnetite hair. Kelso sand is so rich in magnetite that a Texas entrepreneur once staked parts of the dunes as placer claims and planned to separate the magnetite to sell to a steel company. Fortunately for the dunes, this project seems to have languished.

Wind ripples are among the most intriguing features on dune surfaces. Most are a few inches from crest to crest and a fraction of an inch high. They are miniature models of transverse dunes, and like dunes, they are asymmetrical, with a long, gentle windward side and a short, steep lee face. They make interesting patterns around surface irregularities and obstacles, such as bushes. These patterns show that the shape of the surface strongly affects wind currents along the ground, causing the currents to depart considerably from the prevailing direction. At times, one can see active wind ripples on dune surfaces oriented at right angles to the general direction of the wind aloft.

Ripples form when the wind is strong enough, about 25 miles per hour, to raise a good cloud of bouncing sand grains. Impacts of incoming bouncing grains drive the coarser "creeping" grains up the windward slope of the ripple. Some linger in the ripple crest, others tumble down the steep lee face.

A wind blowing 30 miles per hour can drive sand ripples along at a speed of several inches per minute. To demonstrate this, take some toothpicks or medium-sized finishing nails into dunes on a windy day, stick them into several successive ripple crests, and watch the ripples migrate. Another fun experiment is to erase the ripples in a patch of sand two or three feet square by smoothing with your hands and watch the wind make new ones in a matter of minutes.

Long ago, a famous British hydraulic engineer, reasoning from the behavior of ripple marks made by water flowing through laboratory flumes, concluded that strong transverse winds must maintain a powerful eddy over the lee slope of a dune. He suggested that the eddy would undercut the lee side of the dune, making sand avalanche down the slope. Modern experiments with smoke bombs in Kelso Dunes and elsewhere have shown that the idea was wrong. Temporary eddies occasionally form along the lee side of a dune, especially under oblique winds, but no permanent eddy powerful enough to move sand exists there. Under a strong transverse wind, the lee side of a dune is largely becalmed. You can demonstrate this by watching strands of dead grass that you toss onto the lee slope. They move intermittently, usually in a leisurely and aimless way, up, down, or along the lee slope. No determined eddy drives them along.

Dunes provide an unusual ecological niche in the harsh desert environment because they save water. Their porous sand absorbs essentially all the rain that falls on them. After a rain, evaporation from the upper few inches of sand produces a dry layer that insulates the moist sand below from heat and evaporation. This happens because the uniform size of the sand grains minimizes the number of capillary passageways that connect to the surface. Months after any rain, a hole dug several feet into soil adjoining dunes will be bone dry, whereas a hole dug on the windward side of a transverse dune will usually penetrate moist sand within a few inches, at most within a foot or two. At any time of year, strong winds may blow dry sand off the dune surface, exposing moist sand beneath.

The twenty-two clusters of desert willow trees that grow well up on the south face of the Kelso Dunes not far east of their highest point demonstrates the availability of water. Some trees are dead, but many more are living, blossoming, and making seed pods—all signs of good health. Individual tree trunks approach one foot in diameter, and the height can reach 20 feet. The willows grow right out of dune sand, which is hundreds of feet thick. These trees probably established themselves under climatic conditions more moist than the present, but they survive because the dunes hold water.

Burrowing animals, reptiles, and bugs know they have a temperature-controlled system in the dune sand. By burrowing to a chosen depth, they select a comfortable temperature and humidity. One of the most interesting denizens of the dunes is a little lizard that loves to lie burrowed in sand. The pressure of a passing hiker's foot inspires it to boil out of its burrow and take off across the surface like a streak of light. It can disappear before your eyes, either by stopping and holding still—its coloration providing almost perfect camouflage—or, more likely, by burrowing into the sand. The lizard knows the difference between accretion and avalanche sand. When wanting to escape in a hurry, it heads for the lee face of a dune and literally dives into the loose avalanche sand. Children love these astute creatures.

Inspecting dune surfaces in early morning after a windless night reveals a world of tracks etched on the soft sand. A lizard leaves tail streaks between its footprints. A sidewinder rattlesnake leaves a distinctive series of cuspate curves. You have to see a beetle in action to appreciate how its track is made, and that is easy because they are about in the daytime. Tracks of larger animals, such as rabbits, foxes and coyotes also show up well in the sand.

Indians camped near the Kelso Dunes in bygone days. Artifacts, such as arrow points, are still occasionally found and large grinding stones were once abundant in the vicinity. Stones, carried into the dunes and reddened by fire, mark sites where Indians had campfires. They reputedly washed blankets and furs by impregnating them with sand and then shaking them out to remove grease, dirt, and vermin.

Dunes are normally quiet, but occasionally, they break the silence with a deep, low-pitched sound like the moaning of a diesel locomotive far, far away. This happens during or shortly after periods of strong wind. The phenomenon, known as singing dunes, has attracted considerable attention. Scientists attribute the sound to sand avalanching down lee slopes. You can make the noise by walking down the lee face of a dune after a strong wind, starting large sand avalanches.

So, where did all sand come from? Why are the dunes so far out in the Kelso Valley, rather than tucked up against one of the bordering mountain ranges? And, finally, how old are they?

The first question is the easiest. About 35 miles west, the Mojave River flows east from narrow Afton Canyon to build a broad alluvial plain.

Every time it floods, the river renews the supply of raw, loose rock debris, much of it sand, on the surface of this plain. The prevailing westerly winds blow the sand off the alluvial plain, east into the Kelso Dunes. The uncommonly large amount of magnetite in the Kelso Dunes probably comes from Afton Canyon, where mines have produced the mineral in commercial quantities. Strong north winds blowing out of of the valley of dry Soda Lake near Baker probably contribute additional sand.

The location of Kelso Dunes well out in a valley is by no means unique. The huge dune in Eureka Valley and the Death Valley dunes near Stovepipe Wells also lie in valleys. A long study of the behavior of transverse dunes in the Kelso complex provided a possible explanation. It showed that individual transverse dune ridges moved back and forth cumulative distances well in excess of several hundred feet in ten to twelve years, but ended up within a few feet of their initial position. During this same period, the wind removed and redeposited a thickness of several hundred feet of sand on dune crests, with an almost perfect balance between accumulation and erosion. Although the dunes have been very active, they have not moved far in any direction.

The ability of wind to transport sand increases with approximately the cube of its speed, so doubling the wind speed increases its carrying power by a factor of eight. Occasional very strong storm winds from south, north, or east are able to balance the effects of the more prevalent, but usually gentler, winds blowing from the west.

So what caused the dunes to start to grow about where they are now? One possibility is that some perturbation on the valley floor, such as a low, rough outcropping of bedrock, localized the initial accumulation of sand. A seemingly more likely explanation, in view of the fact that other large dune masses also prefer valley floors, is that their location reflects a node condition within the complex of conflicting wind patterns. The surrounding mountain terrain could play a part in creating such a node.

When we ask how old the dunes are, we mean when did great piles of sand start accumulating at this locality? Unfortunately, no datable material has been found within the dunes; they contain no internal evidence of their age. If we knew how much sand is now being added to the dunes every year, and divided that into their total volume, the result might be considered a very crude estimate of their age. But it would not be particularly reliable, because the rate of sand accumulation has almost certainly varied with climatic changes. A better approach is to ask how long the Mojave River has been spreading fresh debris over the alluvial plain at the mouth of Afton Canyon. The river gets its water from the high San Bernardino Mountains. During the last glacial episode, those mountains certainly shed much more water than they now do, so the Mojave River must have been much larger then. We know it supplied most of the water to maintain a large lake in the Manix basin upstream from Afton

Canyon (Vignette 15). This lake is thought to have started overflowing through the Afton Canyon channel roughly 14,000 years ago. That could have begun the creation of the alluvial plain and, subsequently, of the Kelso Dunes.

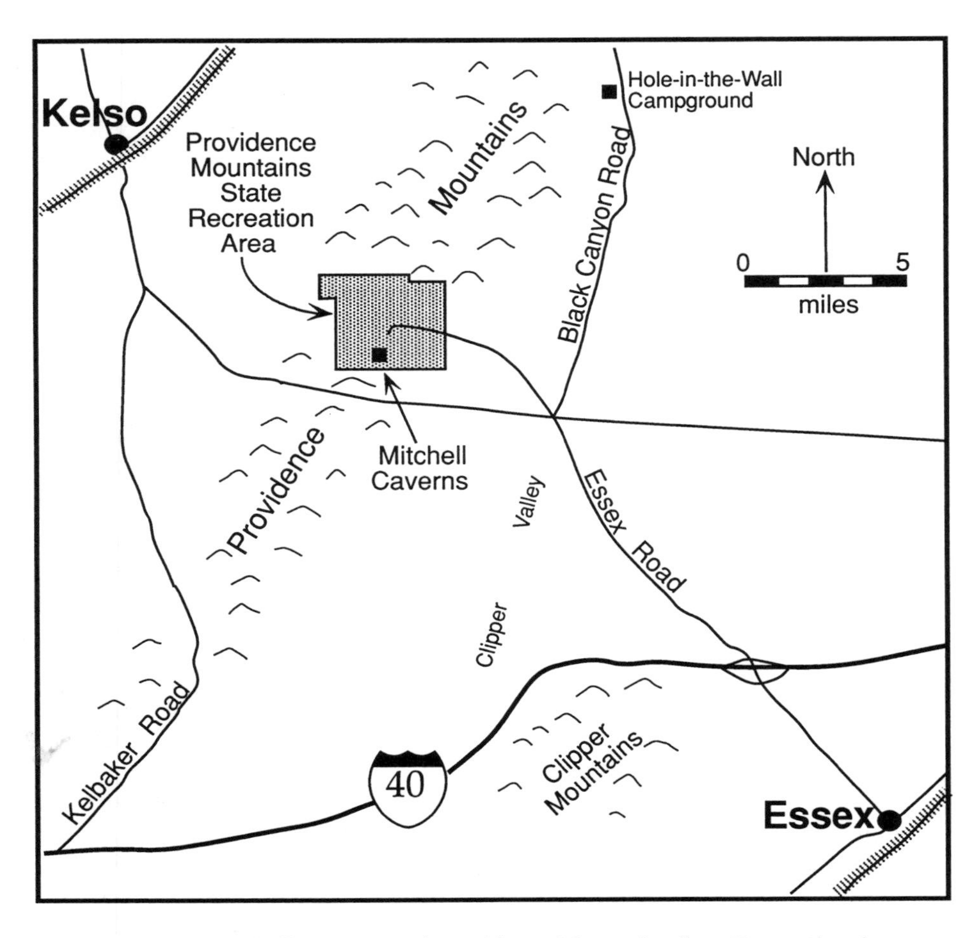

Location of Mitchell Caverns and Providence Mountains State Recreation Area.

Mitchell Caverns are the central attraction of the Providence Mountains State Recreation Area, a 5,900-acre reserve on the eastern slope of the Providence Mountains. Tours of the electrically lighted El Pakiva and Tecopa caverns are conducted by park personnel from mid-September through mid-June. Call the recreation area for a schedule of cave tours. The caves remain at about 65°F year-round, so if the day is warm you may want to bring a sweater for the tour. The U.S. Geological Survey, 7.5-minute, Colton Well and Fountain Peak topographic quadrangles cover the area. To get there via Barstow, drive east on Interstate 40 to Essex Road, 102 miles from the junction of Interstates 40 and 15 at Barstow. Follow Essex Road 9.7 miles to the junction with Black Canyon Road. Bear left at this Y-intersection and continue another 5.9 miles to park headquarters.

VIGNETTE 18

A CALIFORNIA RARITY

Mitchell Caverns

SAN BERNARDINO COUNTY

Limestone caverns are curious, spooky, and awe-inspiring phenomena. It seems a shame that California has so few. The necessary ingredients for cavern formation are fractured carbonate rock and a good supply of water. Carbonate rocks are scarce in California, except in the eastern Mojave Desert. There you will find Mitchell Caverns.

This site, at an elevation of 4,300 feet, offers a wonderful panoramic view of the eastern Mojave Desert. Descriptive plaques identify many of the nearby ranges—and some that are not so nearby. On a clear day, you can see mountains in Arizona, including the Hualapai Mountains, 100 miles away.

Thick layers of volcanic ash cover much of the landscape. The ash erupted ten million years ago in a catastrophic explosion from a caldera in the Woods Mountains about 15 miles northeast of the caverns. The ash buried much of the adjacent countryside, lapped onto the Providence Mountains, and makes up the flat cap of Wildhorse Mesa to the northeast. Careful study of the flanks of Wildhorse Mesa reveals profiles of hills and deep valleys buried beneath the ash. This ash is well exposed at Hole-in-the-Wall Campground on Black Canyon Road, about ten miles north of the intersection with Essex Road.

A broad, one-half-mile trail leads south from park headquarters to the caverns. Rocks along the way tell much of the story of cavern formation, the main subject of this vignette.

The first good outcrops are limestone, part of the Bird Spring formation of Pennsylvanian and Permian age, roughly 250 to 300 million years

View looking northeast from park headquarters of flat-lying tuff of Wildhorse Mesa. Craggy lower slopes of the mesa are older hills that were buried by the tuff of Wildhorse Mesa.

old. A close look at the outcrops reveals that the gray limestone contains bits of orange and brown chert, a microcrystalline variety of quartz. The chert exists in scattered, irregular blobs and in veins lining fractures. This association of chert blobs with limestone is common for rocks of this age; the blobs formed by silica that accumulated with the limy sediments on an ancient ocean floor. The chert stands out in relief on the surface because it resists weathering and erosion more successfully than the limestone. Jagged pieces of chert on the outcrop quickly wear out the soles of your boots and easily tear your clothes (and skin, if you're not careful).

Watch for fossils in the limestone along the trail. (Because this is a state reserve, fossil collecting is prohibited.) The Bird Spring formation contains the remains of quite a variety of animals that lived in sea water during Pennsylvanian and Permian time. The brachiopods look a bit like clams in having a pair of shells, but are not actually related to them. Look for the dark gray cross-sections of their shells exposed in the rock surface.

Other fossils include fusulinids, which are about the size and shape of grains of puffed rice, but are actually the remains of floating, one-celled animals. They abound near the cavern entrance. Crinoids, sometimes called "sea lilies," really are animals, even though they anchor themselves to the sea floor by long stalks made of stacked calcite disks. Look for crinoid stalks that suggest stacked miniature calcite coins in the rock, or for loose stem pieces, about ¼ inch in diameter, that look like little coins with holes through them, like petrified washers.

One-celled animals called fusulinids, typically about one-half inch long, were common in the Bird Spring sea 250 to 300 milllion years ago. Look for their remains in limestone along the trail.

We have found the limestone; now what role does water play in cavern formation? Rain reacts with atmospheric carbon dioxide to make carbonic acid, which makes rainwater slightly acidic. The club soda you buy in the grocery store is a more concentrated solution of carbonic acid. Polluted air normally contains nitrogen and sulfur oxides, which also react with rain to make acids. That is why rain downwind from industrial or heavily populated areas is much more acidic than normal rain—hence the term, "acid rain."

Both limestone and its first cousin dolomite contain carbonate (a carbon atom bonded with three atoms of oxygen: CO_3), which reacts with acids of any kind to form carbon dioxide gas and water. Squirt a bit of acid on limestone, and you can watch it fizz as carbon dioxide escapes and the calcium and magnesium go into solution. The same thing happens in a more subdued way when rain falls on the rock.

Limestone dissolves readily in the abundant rain of wet regions to make valleys. In dry regions, limestone resists erosion and normally stands topographically high as ridges and cliffs. But even there, exposed limestone surfaces dissolve in the occasional rain, becoming rough and pitted. You can easily ruin a pair of nice boots by walking on those surfaces or tear your pants by sitting on them.

Rocks in the high peaks of the Providence Mountains are reddish-brown, unlike the gray limestone at your feet. These are igneous rocks that intruded and deformed the Bird Spring formation during Jurassic time, about 160 million years ago. Rocks downhill from the trail are also igneous, but they were moved against the limestone along faults, instead of intruding it. So the caverns are in a thin wedge of limestone that was faulted against igneous rocks on one side and intruded by them on the

Steep "fins" exposed along the trail to the cavern entrance show that bedding in Bird Spring formation is nearly vertical. View looking north from cavern entrance.

other. Those experiences left the rock badly fractured and vulnerable to the dissolving effects of acidic rainwater. Even the orientation of the original bedding in the limestone is hard to see. In fact, it is nearly vertical. The steep, smooth surfaces in rocks along the trail are bedding surfaces. They, and the many fractures, allow water to move easily into and through the rock, exposing it to dissolution. Boxy networks of chert seams highlight some of the fractures.

Pits and grooves in the limestone near the first diversion culvert show the dissolving effects of slightly acidic rainwater. At the second diversion culvert is a small entrance to a cavern, not open to the public. A little farther along the trail, around a right-hand bend, the entrance to the public caverns appears across a gully. This view shows that the sedimentary bedding, indicated by steep fins on the rock face, tilts down to the east at an angle of about 80 degrees. When you reach a point across the gully from the public cavern entrance, bedding in trailside outcrops is well defined by alignment of chert stringers and by very thin layers in the rock.

Another few steps bring you to a radical change in the color and texture of the rocks above the trail. For several yards, the rock is pale brown, smooth, and contains large circular bands. This is a patch of flowstone, a form of limestone often called travertine. It forms as groundwater deposits calcite on the walls or floor of a cave. Flowstone lines many caverns; this exposure may mark the top of a cave room that was exposed by trail excavation. Is it possible that a large open chamber exists directly under your feet?

About five yards past the end of the wooden bridge, watch for nicely preserved fusulinids, crinoid stems, and brachiopods in the dark gray limestone.

The cavern entrance is a good place to think about how caves develop. They form where groundwater, channeled through fractures in limestone, dissolves large holes in the rock. As a fracture enlarges to a critical width of about one-quarter of an inch, water flow becomes turbulent, greatly enhancing solution of the rock. These large fractures thus enlarge even faster, robbing adjacent fractures of water. A few fractures eventually channel most of the water and develop into caverns.

Speleologists, who make it their business to study caves, think that most cavern enlargement occurs when a cavern lies just below the water table. They point out that many caverns are nearly horizontal, as are water tables, even if bedding in the enclosing rocks is inclined (this is true at

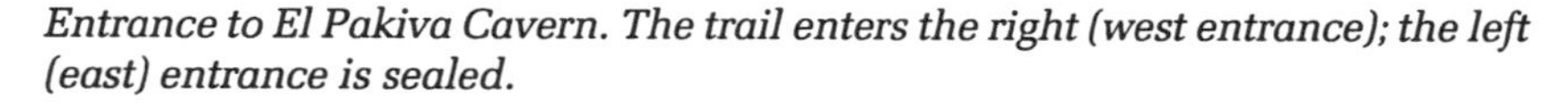

Entrance to El Pakiva Cavern. The trail enters the right (west entrance); the left (east) entrance is sealed.

Dense collection of rapier-like stalactites emerging from cracks in the ceiling. Longest is about 4 feet long.

Tubular stalactites and columns. Longest columns are about 4 feet long.

Mitchell Caverns). Also, the zone just below the water table is a favorable site for dissolution, because when downward-percolating rainwater mixes with groundwater, the resulting mixture is undersaturated in calcite. If the water table drops, caverns begin to fill in with the ornate dripstone deposits that make them so bizarre or attractive, depending on your point of view.

Dripstone deposits form as groundwater seeps from the ceiling or walls of a cavern. Dissolved carbon dioxide in the water makes it acidic enough to dissolve the limestone. If the groundwater dripping into a cave loses some of its dissolved carbon dioxide to the cave atmosphere, it becomes less acidic and must also precipitate part of its load of dissolved calcite, thus forming dripstone deposits.

The most familiar dripstone formations are the stalactites that hang like stone icicles from the ceilings of many caves. They form where water dripping from the ceiling loses carbon dioxide and deposits a film of calcite around the surface of the drop. When the falling drop breaks through that film, it leaves a thin ring of calcite attached to the cave roof. Successive drops add more minute rings of calcite, one after the other, eventually building a thin stalactite that is hollow, like a soda straw. Typically, it takes something like twenty years of fairly steady dripping

Bulbous stalactites and flowstone. Steps for scale.

Cave shield, about three feet across.

to add an inch to the end of a stalactite. If water flowing across the ceiling encounters a soda straw stalactite it will flow down the outside of the tube, enlarging it to the familiar icicle of stone.

Water that drips from the tip of a stalactite and splashes onto the floor will build a mound of calcite, a stalagmite. They form as solid columns rather than hollow tubes. If the stalactite and stalagmite meet, they form a column from floor to ceiling.

Mitchell Caverns is rich in stalactites and stalagmites. It also contains a diverse collection of other deposits, including sheets of flowstone, deposited by water flowing down the cave walls. The origin of the irregular and branching stalactites known as helictites is obscure and a matter of continuing debate among speleologists.

The most prized formation in Mitchell Caverns is a large travertine shield, a semicircular sheet of calcite a few feet across and an inch or two thick. Shields form when water seeps slowly from a crack in the cavern wall, building layers of calcite outward.

Mitchell Caverns are several thousand feet above the floor of Clipper Valley, so the water table is well below the caverns. Why are they here? They probably began to form millions of years ago, when the volcanic rocks seen on Wildhorse Mesa lapped against the Providence Mountains.

At that time, the ground surface was above the level of the caverns, and the water table could have been above the caves. Erosion has since exposed the mountainside. Faulting has also lifted the caverns above the valley floor. Or perhaps the caverns formed after the mountains rose but when rainfall was abundant enough to maintain a through-going flow of groundwater, presumably during the ice ages.

The last great ice age, which ended about 10,000 years ago, was probably the last period when rainfall was heavy enough to enlarge the cave. These days, periods of exceptional rainfall merely dampen the cavern walls but probably do little to enlarge the cave or to add to its dripstone decorations. Most of the year the cave walls are dry.

Map of El Pakiva and Tecopa caverns. —Modified from maps by R. Aalbu (NSS Bulletin, 1990), and by the Southern California Grotto of the National Speleological Society (unpublished)

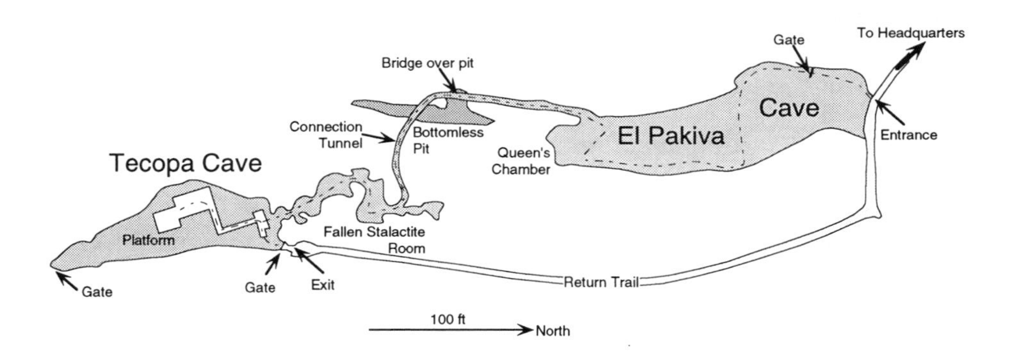

Diagrammatic cross section of El Pakiva and Tecopa caverns. —Modified from maps by R. Aalbu (NSS Bulletin, 1990), and by the Southern California Grotto of the National Speleological Society (unpublished)

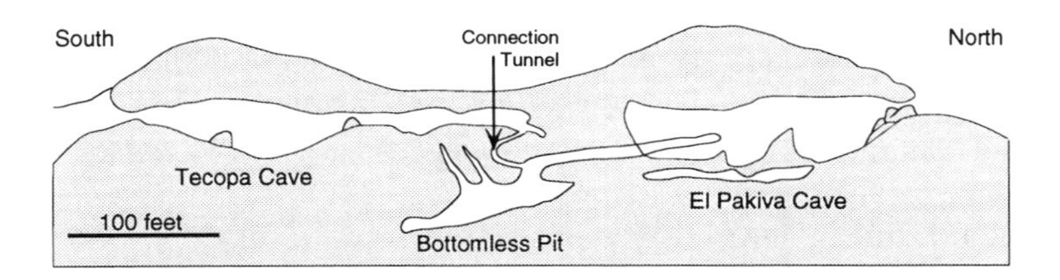

The following notes and maps are meant to supplement the tour provided by park personnel. The guided tour actually goes through two separate caves, El Pakiva and Tecopa, which were joined by a tunnel drilled in 1969-70. Depending on the source, el pakiva translates to "pool of water from the eye of the mountain" or "devil's house." Tecopa was a chief of the Shoshone tribe. El Pakiva is about 200 feet long, and the main chamber of Tecopa is about 150 feet long (see map).

The tour starts at the west entrance of El Pakiva (the adjacent eastern entrance has been walled off). After climbing steps and passing through the gate, the path curves left into a large room marked by large chunks of rock, called breakdown blocks, that have fallen from the walls or ceiling. A sharp right turn brings you into the main chamber of El Pakiva. Here, the ceiling is 40 to 50 feet above you, and the trail hugs the east wall of the cave, a limestone slab that leans out over the flat floor. The opposite (west) wall, on your right, is mostly flowstone.

A walk of about 100 feet brings you to the Queen's Chamber, a magnificent room decorated with thousands of stalactites, stalagmites, and columns. There is much to see here, and your guide will point out the more obvious wonders. The path then bends to the right and climbs out of the Queen's Chamber into a narrow natural passageway that leads to a lower level of El Pakiva. In another 100 feet, the trail comes to a bridge over this lower level, which is known as the Bottomless Pit. Shortly thereafter, the trail enters the artificial tunnel that connects El Pakiva and Tecopa. This tunnel leads to the twisting northern segment of Tecopa and through the aptly named Fallen Stalactite and Hollow Floor rooms, eventually bringing you into the large main chamber of Tecopa. The tour ends on a platform above the dust in the main chamber of Tecopa, where the ranger will tell of the previous inhabitants of this cave, both human and nonhuman.

When visiting Providence Mountains State Recreation Area, be sure to take in the small museum and the Mary Beal Nature Study Trail, a half-mile tour of the flora and fauna of the area. The museum walls are made from an astonishing variety of rocks, all of which Jack Mitchell, the caverns' early developer, claims to have collected from the surrounding mountains. One can spend a pleasant afternoon just admiring these rocks. The museum steps make a great place to gaze out across eastern California and Arizona and ponder California's earlier, wetter history.

VIGNETTE 19

VOLCANIC DEVASTATION

The Bishop Tuff

INYO COUNTY

U.S. 395 in eastern California is a treasure: a straight road that passes through some of the most splendid scenery in the country, from high desert to wooded valleys of the eastern Sierra Nevada. The finest stretch of the highway skirts the eastern foot of the Sierra Nevada, between Little Lake and Bridgeport. North of Little Lake, the Sierra crest rises rapidly, reaching its apex at the 14,495-foot summit of Mt. Whitney, west of Lone Pine. In the shadow of the high Sierra, the highway follows the Owens Valley, climbing from about 2,500 feet in the south to about 4,100 feet at Bishop. The vestigial Owens River—most of which is diverted into the Los Angeles aqueduct—flows along the eastern side of the valley.

Just north of Bishop a large plateau, the Volcanic Tableland, rises from the floor of the Owens Valley. The tableland interrupts the gentle grade of the valley floor, and the road climbs nearly 3,000 feet in ten miles, ascending Sherwin Grade. The base of the grade lies in desert sage country, the top in forests of piñon pine, Jeffrey pine, and juniper, with quaking aspen along streams. The remnant Owens River emerges from a deep gorge cut into this plateau just north of Bishop.

Geologists recognize this area as the site of a geologically recent volcanic eruption of almost unimaginable violence. About 730,000 years ago a colossal eruption blew volcanic ash and other fragmental debris from a large elliptical vent area. The debris from this eruption forms the Volcanic Tableland and is known as the Bishop tuff.

The volume of the Bishop tuff is about 150 cubic miles, enough to cover all of Los Angeles County to a depth of 200 feet. No volcanic

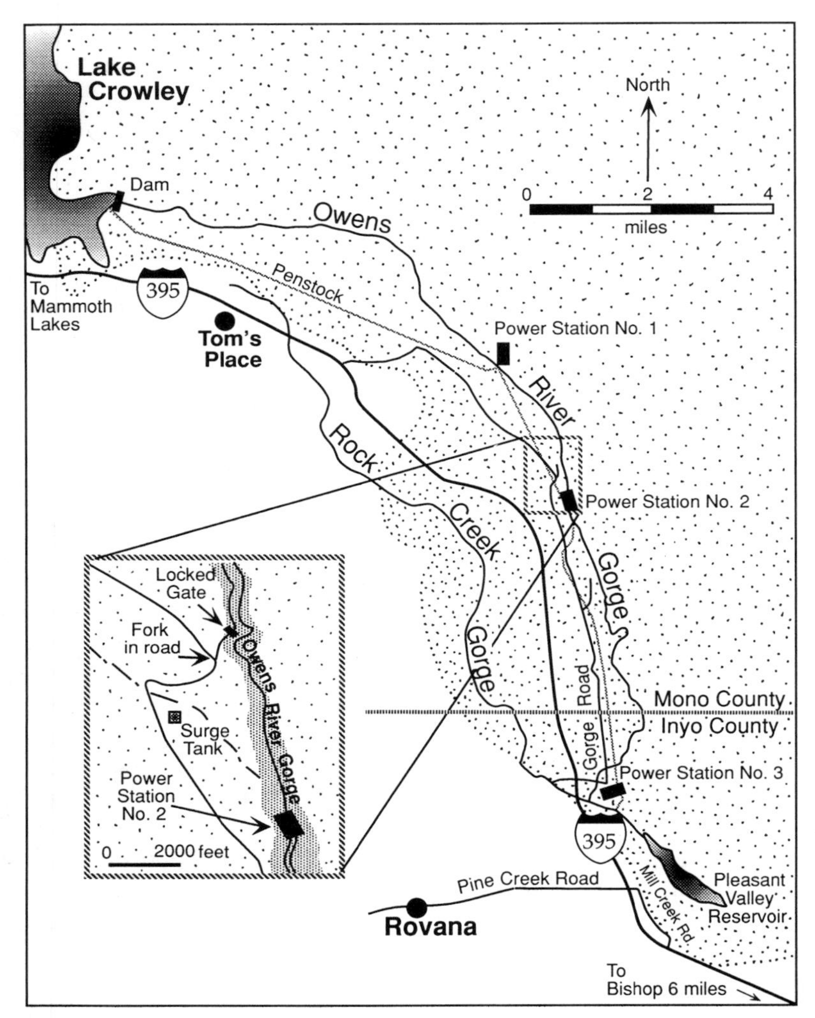

Location of Owens River gorge. Dot pattern shows approximate distribution of Bishop tuff. Inset map gives location of hike down powerhouse road.

The U.S. Geological Survey, 7.5-minute, Rovana topographic quadrangle covers the area. To reach these exposures from the south, drive north along U.S. 395 through Bishop. At the north end of town, the highway makes a 90-degree westward (left-hand) bend; from this bend, drive 11.8 miles to Gorge Road. Look for the turnoff two miles past Mill Creek Road and the turnoff to Rovana. Turn right on Gorge Road and travel 0.7 mile to a T intersection at the Los Angeles aqueduct penstock (large pipeline); turn left and drive north 6.0 miles along a poorly paved road to a fork. Upon reaching the fork, bear right and drive 0.2 mile to a locked gate and park. If the gate is open, park outside or you might get locked in. Travelers coming from the north should look for the turnoff to Gorge Road 11.4 miles south of Tom's Place turnoff. Along Gorge Road from U.S. 395 to the above stop, you will traverse the upper, eroded surface of the Bishop tuff.

eruption in all of recorded human history has even remotely approached that size. Its volume was some 600 times greater than that of the 1980 eruption of Mount St. Helens. Even the great 1912 eruption of Katmai volcano, in a sparsely inhabited part of Alaska, produced only about three to four cubic miles of tephra, fragmental debris produced by a volcanic eruption. Prehistoric eruptions larger than the one that produced the Bishop tuff are known, but they are rare. Several happened in Nevada, Colorado, and Utah between 20 and 40 million years ago. Others built much of the country around Yellowstone National Park between 600,000 and 2 million years ago. More eruptions on the scale of the one that produced the Bishop tuff will probably happen in eastern California and Yellowstone Park in the geologically near future.

One good way to appreciate these enormous volcanic eruptions is to study the Bishop tuff on the Volcanic Tableland. Passable paved and graded roads lead to spectacular exposures in the Owens River gorge.

Conical mounds 50 to 100 feet high dot the surface of the tableland. They mark the sites of old steam vents, called fumaroles, through which volcanic gases escaped (see Vignette 20 on Hot Creek).

The large penstock east of the road carries water from the Los Angeles aqueduct to a hydroelectric powerhouse. The big tanks are surge tanks. They absorb the energy of the first surge of water that comes down the pipe when a valve is opened, preventing damage to the system.

Volcanic tableland as seen from hill west of Bishop, looking north. Casa Diablo Mountain on right; Glass Mountain on left.

Volcanic tableland as seen from Sierra View, in the White Mountains east of Bishop. Steep Wheeler Ridge segment of the Sierra front in background.

Low-angle aerial view, looking south, of fumarolic mounds on eroded surface of Volcanic Tableland. Mounds are typically 10 to 20 feet tall. Shadowed slope in foreground is fault scarp; the Bishop tuff has been stretched considerably in the past 700,000 years as the Owens Valley has widened by faulting.

Owens River gorge and Volcanic Tableland, from the air, looking north. Power station number 3 and north end of Pleasant Valley reservoir are visible at bottom; penstock from power station number 2 is visible at left. Hill in center horizon is Casa Diablo Mountain.

Aerial photo, looking northwest, of upper Rock Creek gorge, upper Owens River gorge, Lake Crowley, and Long Valley caldera. The southern boundary of the caldera runs along the near side of Lake Crowley. —John S. Shelton photo

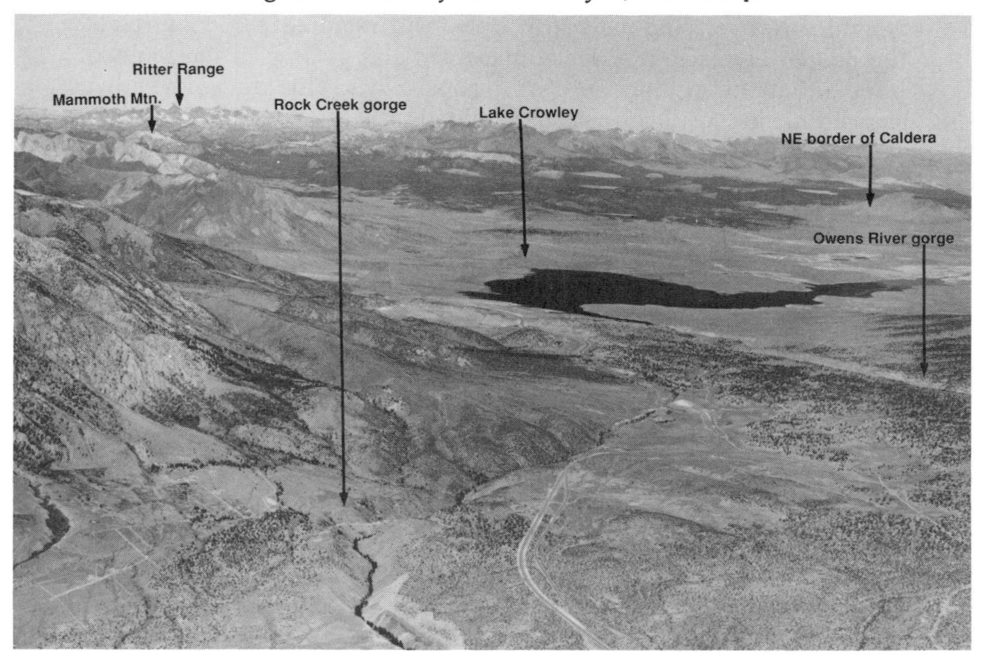

Walk along the powerhouse road past the locked gate, down the west side of the Owens River gorge, which is about 400 feet deep here. All the rocks exposed in the walls of the gorge are Bishop tuff. Look at the rock on the north side of the road, uphill from the old dirt road west of the gate. It is glassy and porous, a pale gray rock that consists of chunks of pumice, or volcanic foam. Deposits like this form when gas-rich magma blows out of the earth, filling the air with blobs of molten volcanic glass. Trapped gases expand the blobs into these bubbly little sponges of pumice. Heft a piece of the rock and note its relative lightness—it is extremely porous because of all the bubbles.

Look closely at the pumice, perhaps with the aid of a hand lens, to see small crystals scattered through the glass foam. Clear ones are quartz and feldspar, and black specks are biotite, an iron-rich mica. You might also see small bits of older rock that were caught up in the magma as it erupted.

The appearance of the tuff changes farther along the road. About 50 feet below the gate, the tuff becomes less porous, with blobs of pale pumice, ranging from a few tenths of an inch to over an inch in diameter, in a pink matrix. Another 100 feet or so farther, the rock is denser still, the pumice clots flattened to dark pink streaks. All the porosity is gone. About halfway down the road, a short distance above the color change from pink to dark gray, the pumice clots are black and flattened. Past the color change, in the lower half of the gorge, the tuff is a uniform dark gray, and the pumice clots become difficult to see. Heft another piece of rock here, and note how much heavier it seems. The bubbles have been flattened, and the tuff has lost its porosity.

These progressive changes reflect the weight of the overlying tuff. The tuff erupted as an effervescing froth of lava that was torn apart by internal gas pressure. When it came to rest, it was a hot, plastic, porous mass. The weight of the overlying material, coupled with residual heat, compressed the deeper layers into a dense mass, remelting some of the blobs of pumice, flattening others, and turning them black. This process is known as welding; the resulting rock is a welded tuff.

You have undoubtedly noticed clusters of little one-inch holes in the rock. Some sort of strange, tuff-boring worm, perhaps? No, the holes were drilled by geologists studying the magnetism of the tuff. These paleomagnetic studies played a key role in plate tectonic theory in the 1960s, because the Bishop tuff shows normal polarity (that is, the direction of magnetism in it is the same as the present-day magnetic field), but rocks only slightly older show reversed magnetism. This reversal, which occurred about 730,000 years ago, is the most recent one in the earth's history. The paleomagnetism and age of the Bishop tuff helped to determine when the reversal occurred.

Look across to the east wall of the gorge. Vertical fractures there divide the rock into columns, making it look like a bundle of pencils. This is a

Spectacular tulip-shaped aggregation of columnar joints in the upper part of the Bishop tuff, on the east wall of the gorge at power station road. Columns are 3 to 5 feet wide.

world class example of columnar jointing. Joints are simply fractures; columnar joints like these form when a rock contracts as it crystallizes. The fractures are generally oriented so that the columns are right angles to the surface of the flow, parallel to the direction of heat flow. The nearby Devil's Postpile is another well-known example of columnar jointing.

Most of the columns in the east wall of the gorge are remarkably regular, with five or six sides, and large, five to ten feet across. At places they curve and converge, like the petals of great stone tulips fanning from their stem, then reaching straight up to the surface. These "flowers" probably formed around fumaroles, when the tuff was still hot. Fumaroles are channels that carry volcanic gases; heat flows outward from them and then up to the surface, just as the radiating columns do. In the lower half of the cliff, in the densely welded, dark gray zone, the columns are as much as 30 to 60 feet across, but they are too crudely formed to be impressive.

Fossil fumaroles stand high on the present surface because the gases that streamed through them deposited minerals that cemented the tuff, making it more resistant to erosion than uncemented tuff in the poorly welded upper part of the deposit. These fumarolic mounds cluster

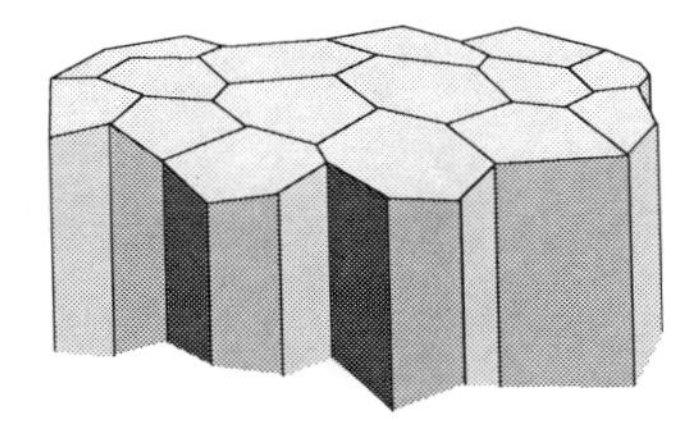

Columnar jointing. When a hot lava or tuff deposit cools, it contracts. Cracks tend to form perpendicular to cooling surfaces (for example, the surface of the deposit) and are most mechanically stable when they intersect in 120° triple junctions. This produces six-sided columns, although five-sided columns are also quite common.

around the gorge, probably because the gases that formed them were mostly steam derived from the ancestral Owens River, which was buried under the red-hot tuff. Similar fumaroles formed above the site of Spirit Lake after the 1980 eruption of Mount St. Helens.

It seems obvious that a river eroded the current gorge, but hardly any water now flows through it. How was it carved? To answer this question, we must start with the origin of the tuff through which the gorge was cut. The Bishop tuff eruption was centered east of Mammoth Mountain and produced a depression about ten miles wide by twenty miles long. When a large volume of molten magma erupts, it leaves the roof of the subterranean magma chamber unsupported. This surface soon collapses, creating a volcanic depression called a caldera. The Long Valley caldera, home of the Bishop tuff, is one of the largest known.

The brief and intensely violent history of the Long Valley region began several million years ago, when molten magma began to collect a few miles beneath the valley floor. The magma began to erupt, producing a cluster of volcanoes, stubby domes of extremely viscous rhyolite magma that rose where the caldera would later form. The only one that survived the climactic eruption is Glass Mountain, a large dome complex of rhyolite and obsidian (rhyolite glass) that formed between one and two million years ago.

The main eruption began about 730,000 years ago, when magma heavily charged with red-hot gases exploded from the top of the magma chamber, erupting the Bishop tuff. Volcanic gases (mainly steam and carbon dioxide) collect at the top of a mass of molten magma. This one contained enough gas to produce a tremendous explosion that blew ash and pumice several miles into the air. Evidence of this early phase of the eruption survives in a thin layer of white ash at the base of the Bishop tuff. As magnificent as this eruption certainly was, it was only a preamble to the truly awesome display that soon followed.

The caldera opened as about 150 cubic miles of Bishop tuff (more than the volume of Mount Shasta) boiled out of the magma chamber, quite possibly in a single, climactic blast. The magma erupted from a set of vents aligned along an arc just inside the margin of the present caldera. About half of the erupting cloud poured across the surrounding countryside as a series of ash flows at a temperature of about 1,500° F. This hurricane of red hot ash spread in all directions at speeds up to 100 miles per hour, devastating everything in its path. One great ash flow poured south down the Owens Valley, past the present site of Big Pine; another spread west, over the crest of the Sierra Nevada, and into the valley of the San Joaquin River. Meanwhile, an enormous cloud of ash blew high into the air, perhaps as high as 25 miles, where winds carried it east, spreading it over most of the western states. Geologists have mapped this ash fall as far away as eastern Nebraska and Kansas.

View, looking south, of Owens River gorge, here about 300 feet deep and carved entirely in Bishop tuff. This is a popular rock climbing spot. Climbers use the flattened pumice disks and horizontal joints as hand- and footholds.

The caldera formed as the Bishop tuff erupted; the roof of the magma chamber fell in as the chamber emptied. When it first formed, the depression was about two miles deep, but the erupting Bishop tuff almost immediately filled about two thirds of that.

Shortly after the eruption, a large lake formed in the remaining caldera depression. Meanwhile, the central part of the caldera was lifted as the remaining magma rose beneath it. Some of this magma reached the surface and was extruded to form rhyolite domes and flows. Hills northwest of Hot Creek (Vignette 20) are domes that formed in this way. Presumably, the magma that rose beneath the caldera after the main eruption did not contain enough gas to inspire it to violence.

Doming of the caldera floor, combined with continual seasonal inflow of water from the nearby mountains, gradually raised the lake's water level. Eventually the water level reached the lake's southeastern margin, the lowest point in the caldera rim, and spilled over, carving the deep channel of the Owens River gorge. Once the lake began overflowing, the river probably cut quite rapidly into the easily eroded Bishop tuff. After the spillway eroded deep enough to drain the lake, the river lost its main water supply. Its flow has since fluctuated, increasing during the wet years of the ice ages, decreasing during the dry periods between ice ages.

Lake Crowley, a reservoir impounded behind a small dam at the head of the gorge, dates back only to 1941, whereas the ancient lake whose site it occupies may have drained as long ago as 100,000 years. The dam at the reservoir is too low to provide enough vertical drop to generate significant power, so much of the water is diverted through tunnels to turbines farther downstream in the gorge. Captured, tamed, and hidden from view—a sorry fate for the once-mighty Owens River.

VIGNETTE 20

AN INTIMATE CONTACT WITH VOLCANIC PHENOMENA

Hot Creek

MONO COUNTY

Geothermal power, theoretically, is clean and cheap. But drawing geothermal energy from the ground is more difficult than you might imagine, and tapping it can adversely affect the environment. Corrosive waters, mineral deposits that plug up pipes, groundwater pollution, and limited productive life make geothermal power risky for developers. Nevertheless, several geothermal areas in California are producing significant power or have good prospects. They include the Geysers in Sonoma County, which supplies part of San Francisco's electricity, and the Salton Sea region.

One of the most pleasant places to experience geothermal phenomena is Hot Creek, a public hot springs near Mammoth Lakes. Hot Creek lies within the 10-by-20-mile Long Valley caldera, a product of a massive volcanic eruption 730,000 years ago (see Vignette 19). That eruption produced some 150 cubic miles of volcanic ejecta, the Bishop tuff, that blanketed the surrounding countryside and spread ash as far east as Nebraska. Quite a lot of molten rock remained in the ground after the eruption. During the past 730,000 years, this material has slowly leaked to the surface, producing many volcanoes in the Mammoth area, including Mammoth Mountain, the Inyo and Mono craters and domes (some of which are less than 700 years old), and the hills around Hot Creek. The hot springs and steam vents at Hot Creek owe their existence to the presence of remnant magma a few miles below, the vestiges of the great magma chamber from which the Bishop tuff erupted.

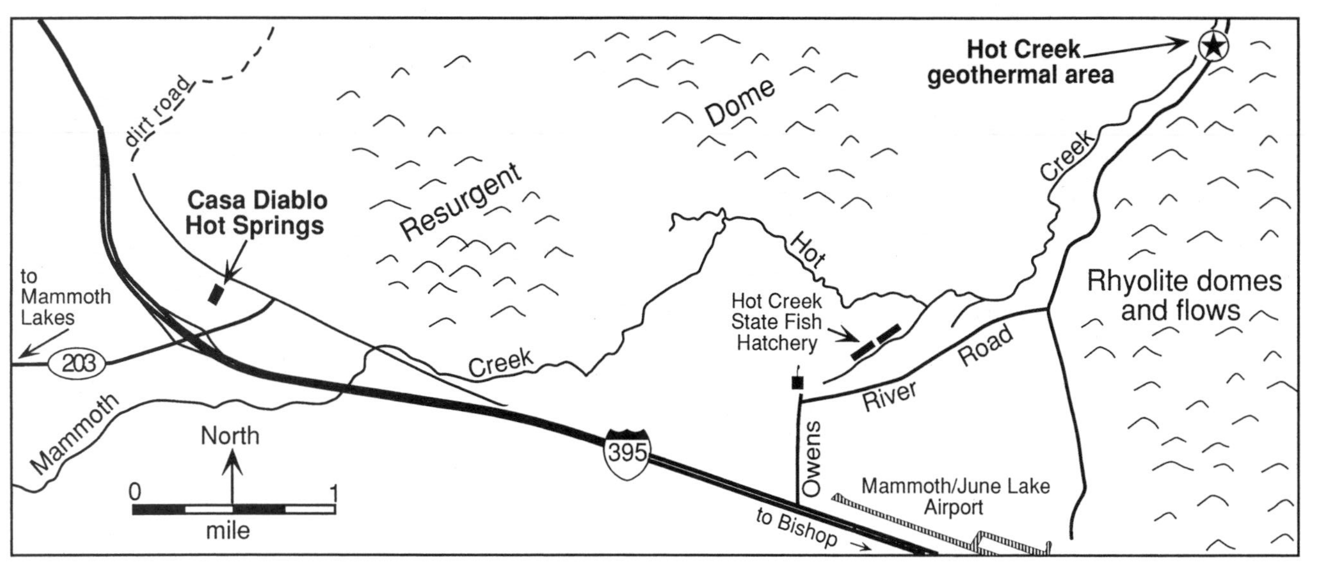

Locations of Hot Creek and Casa Diablo Hot Springs.

Hot Creek is about three miles northeast of Highway 395, between Lake Crowley and Mammoth Lakes. The U.S. Geological Survey, 7.5-minute, Whitmore Hot Springs topographic quadrangle covers the area. To get there, take the road labeled "Owens River Road, Airport" from U.S. 395, just north of the Mammoth Lakes airport, 1.3 miles north of the Convict Lake turnoff and three miles south of the California 203 (Mammoth Lakes) exit. Drive north 0.5 mile and turn east (right); continue another 1.3 miles to a Y-intersection and bear north (left). Drive 1.6 miles to the well-marked parking lot at Hot Creek. Don't turn off at small, unmarked parking areas provided for fishermen. (The creek is popular with anglers who fly cast for trout.) At the Hot Creek parking lot are pit toilets and a shelter for changing clothes.

To get to Casa Diablo Hot Springs, retrace your route to U.S. 395 and drive north three miles to the California 203 exit. Turn right at California 203 and go 0.3 mile to a T-intersection with old U.S. 395, and park along the shoulder of the road. The geothermal area is marked by steaming fumaroles and some buildings.

Aerial view, looking south from above Mono Lake, of Mono and Inyo craters, Long Valley caldera, the resurgent dome, and the crest of the high Sierra. Hot Creek lies south of the hills of the resurgent dome. —John S. Shelton photo

The parking lot at the hot springs is a good place to scout the surrounding landscape. For orientation, the creek flows to the northeast in the gorge below you, and the overlook faces approximately west. Hills west and northwest of Hot Creek are part of the resurgent dome of Long Valley, a mass of magma that welled up into the center of the caldera shortly after the eruption. The White Mountains, capped by lofty White Mountain Peak (14,246 feet), dominate the distant eastern skyline. To the north and northeast, the great mass of Glass Mountain forms the northeastern rim of the caldera. To the west, the top of Mammoth Mountain peeks over the resurgent dome. Spectacular summits to the southwest and south surround Convict Lake; the prominent pointed gray monolith almost due south of the parking area is Mt. Morrison (12,268 feet); to its right is Laurel Mountain (11,812 feet). To the west of these peaks, in the Mammoth Lakes basin, lie the headwaters of Mammoth Creek; its name changes to Hot Creek east of U.S. 395.

Hot Creek flows within a narrow gorge cut deeply into the relatively soft volcanic rocks of the Long Valley caldera. As you peer over the edge, you can see fumes from several volcanic-gas vents, known as fumaroles, most of which are behind protective barriers. You will also see a pool in the creek just downstream from the footbridge, where several hot springs rise from the creek bottom and bathers gather. Bathers generally arrange themselves in an arc along the boundary between the very hot water above the vents and the icy waters of the creek. If you watch for a while, you may see the position of the arc shift in response to fluctuations in cold stream currents and underwater hot spring activity.

As you walk along the short trail to the hot springs, look at the volcanic rocks, chiefly rhyolite, in the walls of the gorge. They are much altered by hot water and volcanic gases, mostly steam and carbon dioxide, along with small concentrations of sulfur compounds. The sulfur gases, mainly sulfur dioxide and hydrogen sulfide, give many hot springs their unpleasant, even infernal, smell—not bad at Hot Creek. Sulfur oxide gases also combine with water to form sulfuric acid, which is highly corrosive to rocks, vegetation, and geothermal pipes. Acid attack has left the rocks around Hot Creek bleached and punky, largely converted to clay. An open pit mine in an inactive geothermal area two miles northwest of Hot Creek produces kaolinite, a white clay used as a filler in products such as paint, plastic, paper, and milk shakes. Not all the rock has been

View south along Hot Creek road to Mt. Morrison (12,268 feet) on the left and Laurel Mountain (11,812 feet) on the right.

Hot Creek bathing area, from the parking area. Many hot water vents surface just downstream (right) of the footbridge. Several of the fumaroles on the far side of the creek formed in the 1980s.

Hot Creek bathing area, looking upstream. Laurel Mountain in background.

Altered obsidian with wavy flow banding, along the path to the bathing area.

completely altered; in places you can recognize remnants of volcanic glass, obsidian, complete with attractive flow banding. The rocks at the kaolinite mine provide a preview of what the rocks at Hot Creek will look like if they continue to be subjected to geothermal activity.

Many hot springs surround themselves with deposits of siliceous sinter, a rock made of a form of silica similar to opal, and travertine, a banded form of calcite (calcium carbonate). Sinter and travertine form when those substances come out of solution in hot water as it ascends, cools, and loses its gases. These minerals also deposit in pipes, creating what we might call geo-atherosclerosis and giving headaches to the geothermal engineers. Sinter is typically white, but attractive blue varieties are locally common. White sinter and travertine abound along the upper part of the Hot Creek trail, marking sites of extinct hot springs.

Partway down the trail, a short wooden walkway leads to a group of vigorously steaming fumaroles. They appeared abruptly during the night of August 24, 1973, about a half day before a small earthquake hit near Bishop. Two small geysers and several smaller vents also formed here at that time, temporarily spouting steam, mud, and hot water. Other new vents formed north of the creek during the four magnitude 6 earthquakes of May 25-27, 1980, which were centered a few miles west of Hot Creek. These events show that earthquakes affect the plumbing system of hot springs. And changes in the vent system may, under some circumstances, forecast an earthquake.

The main objective for most visitors to Hot Creek is a dip in the hot springs. Several signs warn against such activity, but a swim is safe if you follow a few precautions. Be careful around vents, both on the ground and underwater. They can severely scald you, and so can bubbling mud along the bank, as well as the hot sand a few inches below the creek bed. Be forewarned that sulfurous gases will quickly tarnish silver jewelry.

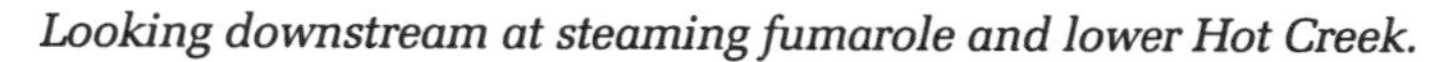

Looking downstream at steaming fumarole and lower Hot Creek.

New fumarole, about 6 feet across, that opened at the northwestern end of the footbridge in early 1991. This area was formerly a beach along the creek.

In the spring, when snowmelt is high, the hot water is so diluted that it feels barely warm. At such times, the creek water is frigid. In the fall, when runoff is low and warmer, the spring waters can be uncomfortably hot. Water level near the bridge varies by several feet, from season to season.

Sitting or swimming near a vent is a perfect occasion to contemplate the deeper meanings of volcanism. While you relax and get pruney fingers and toes, you can watch bubble trails carry volcanic gases from vents on the creek floor, and feel a deep, rhythmic thumping that is probably caused by gas bubbles collapsing in the plumbing below. After emerging rubber-bodied from the hot springs, you can stop at another geothermal site in the Mammoth area: Casa Diablo Hot Springs, just north of the U.S. 395-California 203 intersection.

Fumaroles at Casa Diablo Hot Springs have been a spa for centuries, ever since the Paiute Indians set up bathhouses there. A long succession of stagecoach stops, trading posts, cafes, and gas stations has occupied the spot. Today generating stations tap the geothermal power.

Active fumarole at Casa Diablo Hot Springs. Bleached white, punky alteration is typical of rocks that have been attacked by geothermal fluids.

Geothermal development at Casa Diablo Hot springs as seen from old Highway 395, looking south.

The hillside just east of Casa Diablo marks one of the many north- to northwest-striking faults that cut the resurgent dome. These faults channel and localize geothermal flow. The kaolinite mine northwest of Hot Creek lies along one of these faults. As is quite evident, the rhyolites around here are in a sorry state, altered almost beyond recognition by the hot geothermal fluids that still bubble up through the mud. Take care in viewing the many fumaroles that lie near the road, and do not cross any fences.

Geothermal activity at Casa Diablo has been erratic. Generally, the only surface manifestation consists of small, steaming fumaroles similar to those at Hot Creek. On occasion, as in 1937 and 1959, geysers have spouted hot water and steam as much as 80 feet into the air. On both sides of the road, just north of the power plant, you can see small fumaroles and bubbling mudpots.

Geothermal development has been sporadic at Casa Diablo, since about 1960. Early attempts at power production failed for several reasons: the steam is highly corrosive to pipes, drilling caused geothermal activity to decrease at Casa Diablo, and the hot water that was brought to the surface polluted the groundwater with boron and arsenic. Because environmental regulations forbid the release of a large steam plume, steam that has passed through turbines must be condensed in the large, open building on the east side of the road. Nevertheless, the plant is currently supplying power to the Southern California Edison power grid. Three power stations (the last two dedicated in 1991), with a total capacity of 40 megawatts, now feed power into the California grid. That is enough power to supply the needs of over 10,000 homes.

Glossary

Aa. Lava with a blocky, jagged, clinkery, spinose surface.

Abrasion platform. See Wave-cut platform.

Acidic. Igneous rocks containing more than 60 percent silica.

Accretion sand. Firmly packed windblown sand accumulated under a curtain of hopping (saltating) particles.

Agglutinate. A welded pyroclastic product consisting of formerly pasty lava globs.

Alluvial apron. Smooth deposit of alluvium, of regional extent, sloping gently outward from the base of a mountain face.

Alluvial fan. Fan-shaped deposit of alluvium bordering the base of a steep slope at the mouth of a canyon.

Alluvium. Unconsolidated gravel, sand, and finer rock debris deposited principally by running water; adjective: alluvial.

Amino acids. Organic compounds that are essential units within proteins.

Amorphous. Said of a mineral or other substance that lacks crystalline structure.

Amplitude. Half the height of a wave crest above the adjacent trough.

Angle of repose. Maximum angle of slope, measured from horizontal, assumed by a pile of loose, cohesionless (dry) particles.

Angular unconformity. A break or interruption of the geologic record within layered rocks, in which the upper sequence rests upon an erosional surface that angularly truncates the lower sequence.

Anticline. A fold in layered rocks, convex upward, with older rocks toward the core.

Arroyo. A relatively broad, flat-floored, steep-banked, channel of an intermittent stream in arid or semiarid country.

Ash. See Volcanic ash.

Avalanche sand. Loosely packed, windblown sand that has flowed as an avalanche down the lee face of a dune.

Axis. As used here, the center line of a fold in rocks.

Badland. Barren, steep, intricately dissected terrain developed by erosion in fine but coherent sediments.

Bank (oceanographic). Shallowly submerged, usually sandy, embankment along a sea coast.

Basalt. Fine-grained, dark, primarily extrusive igneous rock, relatively rich in calcium, iron, and magnesium and relatively poor in silicon.

Basic. Igneous rocks with a relatively low silica content, 52 percent or less.

Beach. A constantly changing deposit of loose sand or stones along the shoreline of a water body.

Bedding. The layered structure of sedimentary rocks.

Bedrock. Relatively solid rock, exposed or underlying a mantle of loose rock detritus.

Bluff. High bank, cliff, or headland rising above a flat.

Bomb. See volcanic bomb.

Boulder. A rock fragment larger than ten inches diameter, usually worn and at least partly rounded.

Brachiopod. Clam-like solitary marine invertebrate whose shell pieces are bilaterally symmetrical.

Brackish. Water with a salinity between that of fresh and sea water.

Breakaway scarp. The steep face left at the top of a landslide scar on a hillside.

Breakwater. An offshore structure designed to protect an anchorage from waves.

Breccia. A rock consisting of angular rock fragments held together by a mineral cement or in a fine-grained matrix.

Brink. The sharp breakoff edge between a gentle and a much steeper slope.

Bulrushes. Tall, single-stemmed marsh plants, such as cattails.

Calcite. Widespread, abundant mineral composed of calcium carbonate ($CaCO_3$); the major component of limestone and marble.

Caldera. Large, circular or oval basin formed by collapse following a voluminous volcanic eruption.

Capillary. Tubelike opening so small that it holds water despite the pull of gravity.

Carbon 14. The radioactive isotope of carbon which disintegrates with a half-life of 5568 ± 30 years.

Carbonate rock. Rock composed of the minerals calcite or dolomite, both of which contain carbonate (CO_3); typical examples are limestone and marble.

Capture. See Stream capture.

Cave. Naturally formed subterranean chamber, large enough for a person to enter.

Cavern. A large cave or cave system.

Chaparral. Thicketlike vegetative complex of scrubby bushes densely cloaking areas in semi-arid regions.

Chert. Hard, dense, dull to partly glassy, sedimentary rock composed of microcrystalline silica.

Cinder cone. Cone-shaped accumulation of volcanic cinders erupted from a central vent; see Volcanic cinders.

Clast. An individual fragment large enough to be visible to the naked eye in sediment or sedimentary rock.

Clay. Rock or mineral particles smaller than 0.00016 inch, or crystals of a clay mineral; plastic when wet.

Clastic dike. Tabular intrusive body of sedimentary detritus (commonly sandstone).

Cleavage. Ability of many minerals to split along specific atomic planes.

Cliff. A high precipice in rock, near-vertical or overhanging.

Cobble. Rock fragment, usually worn to rounded, between 2.5 and 10 inches diameter.

Conglomerate. Sedimentary rock consisting of pebbles, cobbles, or boulders, cemented within a sandy matrix.

Continental borderland. The marginal zone of a continent, shallowly submerged by the ocean and underlain by normal continental rocks and structures.

Core drilling. Drilling by a rotary bit that leaves a cylinder of rock within the drill stem.

Country rock. That intruded by an igneous body or enclosing a mineral deposit.

Creep. Slow continuous movement of rock detritus downslope by gravity and processes disturbing the particles; see also Impact creep.

Crinoid. Otherwise known as a sea lily; a bottom-dwelling animal with a slender column supporting a globular, pentagonal body from which appendages extend. Generally, only cylindrical stem segments are preserved.

Crop out. See Outcrop.

Crystal. A many-faced solid bounded by smooth planar surfaces that reflect an orderly internal arrangement of atoms.

Cube. In this context, a value increased by multiplying itself three times.

Daughter product. Chemical element produced by radioactive disintegration of a parent substance.

Debris. Surficial, loose, broken rock particles.

Deep. An obvious depression of the ocean floor.

Desert varnish. See Rock varnish.

Detritus. Loose, disintegrated particles of rock.

Dike. Tabular, igneous, intrusive body, discordant with structure of country rock.

Dip. The inclination from horizontal of any planar surface within rocks, as measured in the steepest direction (e.g., the direction a marble would roll down the surface).

Disconformity. An unconformity in which the beds above and below an erosion surface are concordant.

Disintegrated granite. Rock detritus, formed by break up under weathering, of coarse-grained igneous rocks.

Dolomite. A carbonate mineral with the formula $CaMg(CO_3)_2$; term is also applied to a rock consisting of dolomite.

Domino breccia. A shattered rock in which domino-size fragments are juxtaposed.

Drilling mud. Highly liquid mud, circulated in oil wells, to confine gas and to help bring up drill cuttings.

Dripstone. Cave formation, generally of calcite, formed by dripping or seeping water.

Earthquake. Vibrations within the earth's crust produced by a sudden release of accumulated strain.

Ecology. Relationships between organisms and their environment; adjectives: ecologic or ecological.

Eddy. A circular movement of water or wind, usually in a different direction from that of the main current.

Effective sand transport. Resultant direction of principal sand movement under multidirectional wind conditions.

Electron. Negatively charged constituent of atoms; 1,800 times lighter than a proton or neutron.

Eolian. Pertaining to the wind (for example, eolian deposition of sand).

Erosion. Removal of rock material by any natural process.

Estuary. Shoreline inlet formed by drowning of a river valley.

Fault. Fracture along which blocks of the earth's crust have slipped past each other.

Feldspar. Group of common rock-forming minerals composed principally of silica, aluminum, and oxygen, plus one or more of the elements calcium, sodium, and potassium.

Fission tracks. Path of radiation damage in minerals or obsidian caused by nuclear particles.

Flowstone. Cave formation, generally of calcite, formed by flowing water.

Fluvial. Features of erosion or deposition created by running water.

Focus. Of an earthquake; that point deep within the earth's crust where the shock was initiated. Synonym: hypocenter.

Foliation. Mineralogical or textural banding in rocks, formed primarily by solid-state metamorphism.

Formation. Geologically, a rock body of considerable areal extent with consistent characteristics that permit it to be recognized, mapped, and usually named.

Fumarole. Small volcanic vent that emits hot vapors.

Furnace marble. Metamorphosed limestone formation of probable Mississippian age exposed on Blackhawk Mountain in San Bernardino Mountains.

Fusulinid. One-celled, floating marine animal with a cucumber-shaped, chambered shell up to an inch in length.

Gamma ray. High-energy form of electromagnetic energy, such as light or x-rays, emitted during some types of radioactive decay.

Geothermal. Involving heat from within the earth.

Geothermal power. Energy derived from the internal heat of the earth.

Geyser. Hot spring that intermittently erupts jets of hot water and steam.

Glacier. A body of natural, land-borne ice that flows.

Glaucophane schist. Thinly foliated metamorphic rock containing glaucophane, a sodium-rich iron-magnesium-aluminum silicate, which gives the rock a peculiar blue to lavender coloration; commonly called blueschist.

Grain flow. Transport involving a flowing, liquified, cohesionless mass of particles, commonly sand grains.

Greenstone. Compact, metamorphosed, igneous rock, rich in greenish minerals.

Groin. Narrow jetty of large stones or pilings extending out from shore, designed to capture longshore-drifted sand.

Groundmass. Fine-grained part of a rock (usually igneous) in which larger particles or crystals are imbedded.

Groundwater. Fills pores and other openings to a condition of saturation in subsurface rocks and sediment.

Gruss. The fragmental product of granular disintegration of coarse-grained igneous rock; see Disintegrated granite.

Hanging wall. The side of an inclined fault plane that would overhang if the walls were separated.

Half-life. Time required for one half of any amount of a radioactive substance to disintegrate.

Helictite. Stalactite-like, irregular, curving dripstone deposits.

Ice Age. A period in earth history when large sheets of ice inundated parts of non-polar continents.

Igneous. Rocks formed by the crystallization of molten material (magma).

Impact creep. Discontinuous movement of particles on the ground under impact by saltating grains.

Isotope. Species of an element defined by the number of neutrons in its nucleus; adjective: isotopic.

Isotopic dating. Determining the age of a geological sample by measuring isotopic ratios.

Jadeite. Hard, lustrous, usually greenish mineral; a sodium, iron, aluminium silicate, used for ornamental purposes.

Joint. Planar fracture in a rock without displacement; often in parallel sets.

Joshua tree. Angular, spiny yucca tree of the agave family that grows in clumps in deserts. Named from fanciful resemblance to a gesticulating prophet.

Jurassic. A period of geologic time, 144 to 208 million years ago.

Kaolinite. Clay mineral rich in aluminum and silicon, commonly formed by alteration of feldspars.

Lacustrine. Features or deposits formed in association with a lake.

Lagoon. Shallow, shoreline body of seawater separated from the open sea by an offshore bar or reef; adjective: lagoonal.

Lava. Extruded magma or the solidified product of such.

Lee-side eddy. Circular movement within a fluid current to lee of an obstacle.

Lichen. Plant common on rock surfaces, consisting of a symbiotic combination of an alga and a fungus.

Limestone. Sedimentary rock composed largely of the mineral calcite.

Longitudinal dune. Long, narrow dune ridge parallel to prevailing wind.

Longshore drift. Movement of material, mostly sand, parallel to shore of sea or lake owing to oblique approach of waves.

Magnetite. Black, magnetic mineral composed of oxides of iron.

Magnitude. For earthquakes: a measure of strain energy released during an event. Magnitude is measured on a logarithmic scale, with each increase of one unit of magnitude corresponding to a ten-fold increase in amplitude of ground shaking and a thirty-fold increase in the energy released.

Marble. Metamorphosed limestone or dolomite, usually coarsely recrystallized.

Magma. Molten rock within the earth.

Matrix. Fine-grained rock or mineral particles filling spaces between coarser constituents of a sedimentary rock.

Metamorphic rock. A rock that has undergone sufficient solid-state physical changes by heat, pressure, and stress to be distinct from the parent rock.

Metate. A flattened or concave stone on which Native Americans ground grains and seeds; pronounced meh-tah-tay.

Mica. Group of silicate minerals with perfect sheetlike cleavage.

Mica schist. Metamorphic rock rich in mica and thinly foliated.

Microcrystalline. The texture of a rock consisting of crystals that are too small to be recognized or distinguished under an ordinary microscope.

Miocene. An epoch in the Tertiary Period, from 5 to 23.5 million years ago.

Mineral. Homogeneous, naturally occurring, inorganic, solid substance of specific chemical composition and physical properties.

Mississippian. Geologic period within the Paleozoic Era covering interval 320 to 360 million years ago.

Monolithologic. Sedimentary rock consisting of one species of mineral or particles of one parent rock.

Mudflow. A mass movement of highly fluid muddy debris.

Mudpot. Shallow hot-spring pit filled with steaming, bubbling mud.

Mudstone. Fine-grained sedimentary rock of silt and clay, coarser-grained and more massive than shale; indurated mud.

Narrows. Topographically, a gorge or restricted passage along a stream or through a pass.

Neutron. Electrically neutral particle in an atom's nucleus; equivalent in mass to a proton and 1,800 times heavier than an electron.

Niche. Shallow cave or reentrant in a rock face formed mostly by differential weathering.

Nodular. As used here, a roughly spherical knot of mineral material within a rock, commonly a mica schist.

Normal fault. An inclined fault on which the hanging wall has moved relatively downward.

North American plate. One of eight huge moving plates that make up the outer, solid part of the earth.

Nucleus. Extremely compact and dense core of an atom, around which electrons orbit; composed of protons and neutrons.

Offshore bar. A bar, usually sand, thats surface is above or below the water level, offshore from and parallel to a shoreline.

Obsidian. Volcanic glass; lava that cooled too quickly to form crystals; commonly rich in silicon.

Opal. Fine-grained silica, or solid silica gel, containing water of crystallization giving opalescence.

Ordovician. An era of geologic time, from 438 to 505 million years ago.

Oreodont. Heavily built, short-legged, piglike animal with four-toed hoofs.

Outcrop. An exposure of bedrock at the surface. The rock is said to "crop out."

Pacific plate. One of earth's major plates, lying west of the North American plate, and consisting of the crust and upper mantle under the Pacific Ocean.

Pahoehoe. Basaltic lava with a relatively smooth, undulating surface of small-scale bulbous, ropy, corded, draperylike features.

Paleo-. Prefix connoting great age.

Paleomagnetism. Remnant magnetism established in a rock by earth's magnetic field when the rock solidified or was deposited.

Paleontology. Study of ancient life, largely by means of fossils.

Patina. A colored film on a rock surface produced by weathering (a tanned rock). See Rock varnish.

Pebbles. Small stones, usually worn to rounded, between 0.17 and 2.5 inches diameter.

Peccaries. Piglike animals with sharp tusks.

Pennsylvanian. An era of geologic time, from 286 to 320 million years ago.

Penstock. A pipe for carrying water.

Percolation. Slow movement of a fluid through a porous material.

Permian. An era of geologic time, from 245 to 286 million years ago.

Plate. In a planetary sense, one of eight large drifting plates composing the earth's solid outer part. Plates are roughly 60 miles thick.

Plate tectonics. Movement and deformation caused by interaction of large, drifting, planetary plates.

Plug (igneous). Relatively small, cylindrical, intrusive body filling a volcanic vent. Synonym: volcanic neck.

Plunge. Inclination from horizontal of the central or axial line of a fold.

Pluvial. Cooler, moister conditions in arid or semi-arid areas, possibly coincident with glacial conditions in other regions.

Polarity. In geomagnetism, the direction, north or south, in which a freely suspended magnetic needle points.

Postglacial. The interval following disappearance of continental glaciers from non-polar regions.

Potassium-argon dating. Method of age determination based on decay of potassium-40 to argon-40.

Proton. Positively charged particle in an atom's nucleus; equivalent in mass to a neutron and 1,800 times heavier than an electron.

Pyroclastic. Clastic rock material formed by volcanic explosion.

Quartz. Common rock-forming mineral that is hard, chemically resistant, and composed of silicon and oxygen (SiO_2).

Quartzite. Principally a metamorphic rock formed by some recrystallization of quartz-rich sandstone. Some silica-cemented sandstones included.

Quasi-equilibrium. A near or temporary equilibrium.

Radical. In chemistry, a group of atoms that remains bound together during chemical reactions and behaves like a single atom.

Radioactive. A substance undergoing spontaneous nuclear change (decay) through emission of gamma rays or charged particles.

Reach. A straight or continuous stretch of a stream, river, or shoreline.

Refraction. Deflection of rays and wave fronts in passing between media of different velocities.

Regressive shoreline. One receding from land because of uplift or the fall of sea level.

Reverse fault. One steeply inclined, with relative upward movement of the hanging wall.

Rhyolite. Extrusive igneous rock (lava) of granitic composition, commonly light colored or reddish. Relatively rich in silicon and poor in iron, magnesium, and calcium.

Right-lateral fault. One with sideways displacement and the opposite block moving to the right.

Rill. A shoestring-like, usually ephemeral channel eroded in soft material on steep slopes by small streamlets of water, in parallel sets.

Riser. Steeply inclined slope or cliff between treads (flats) of natural, steplike, topographic features.

Rock. Any consolidated aggregation of minerals or natural glasses.

Rock varnish. Patina or thin coating of dark material abnormally rich in iron and manganese on the exposed surface of a rock.

Saltation. Transport of particles by wind or water in a hopping mode.

San Andreas fault. Southern California's largest fault, along the junction of the Pacific and North American plates.

Sandstone. Sedimentary rock composed primarily of sand-sized particles of rock or mineral, 0.0025 to 0.08 inch in diameter.

San Jacinto fault. Southern California's second largest and historically most active fault.

Sapping. An erosional process commonly involving groundwater seepage, which undercuts the base of a steep face or slope.

Scarp. Linear steep face from a few to thousands of feet high.

Scour channel. Large erosional groovelike channel cut into older material, commonly filled by younger sediment.

Sea arch. An arch-shaped opening through a projecting headland, produced mostly by wave erosion.

Sediment. Unconsolidated particulate matter deposited by some agency of transport, for example, transported by wind or water.

Sedimentation. The process of deriving, transporting, and depositing sediment.

Sedimentary rock. Consolidated and usually cemented sediment, characterized by layering.

Shield. Shelf-like, semicircular dripstone deposit.

Silica. Silicon dioxide (SiO_2).

Siliceous. Rich in silica.

Sill. Tabular igneous intrusive body conformable with layering in host rock.

Silt. Fine particulate rock and mineral matter, dust sized (finer than sand, coarser than clay), between 0.00016 and 0.0025 inch diameter.

Sinter. Hard incrustation of mineral matter precipitated from hot or cold water may be siliceous (SiO_2) or calcareous ($CaCO_3$).

Sorting. Arrangement of particles by size; adjective: sorted.

Speleothem. Any deposit formed by dripping or flowing water within a cavern.

Stalactite. Dripstone deposit built downward by water dripping from cave ceiling.

Stalagmite. Dripstone deposit built up from cave floor by water dripping from ceiling.

Stratification. Layering in sedimentary rocks.

Stratum. Tabular layer in a sedimentary sequence (pl. strata).

Stream capture. Diversion of the headwaters of a stream into a neighboring channel by erosion.

Striations. Parallel scratch marks on a rock surface, usually linear.

Strike. Compass bearing of a horizontal line on the face of an inclined plane.

Syncline. Downfold in layered rocks with younger beds toward the core, limbs inclined inward.

Swash. Sheet of water surging up a beach, generated by a breaking wave.

Tectonic. Forces involved in deformation of the earth's crust.

Tephra. Fragmental products of volcanic explosions.

Terrace. Steplike landform consisting of a flat tread and a steep riser, commonly of fluvial, lacustrine, or marine origin.

Terrane. Large region underlain by rocks of similar character and history.

Terrain. Tract or region with characteristic physical features.

Terrestrial. Said of a sedimentary deposit laid down on land above tidal levels.

Thrust fault. Gently inclined reverse fault along which one block is thrust over the other.

Torrey pine. Southern California native five-needle Pinus, species *torreana*, needles 7 to 10 inches, grayish to dark green, 30 to 40 feet tall, brown cones 4 to 6 inches.

Traction. Mode of transportation that moves particles across a surface by rolling or sliding; wind or water usually involved.

Transgressive shoreline. Moving inland owing to submergence by rising sea level or sinking land.

Transverse dune. With its crest transverse to the wind direction.

Transverse Ranges. Geological province of southern California with east-west structural grain, thus transverse to the normal north-northwest trend.

Travertine. Accumulation of calcium carbonate resulting from deposition by ground or surface water.

Tread. Flat element of a steplike form.

Tuff. Rock formed of consolidated tephra.

Tuff breccia. Pyroclastic rock consisting of both fine and coarser angular fragments, usually of rock.

Turbulence. State of high disorder, within flowing liquids, with rapid changes in velocity and direction of flow; a swirling movement.

Unconformity. A surface of erosion separating younger deposits from older rocks. See Angular unconformity and Disconformity.

Uranium. Radioactive chemical element that spontaneously decays to other elements, some stable, some radioactive.

Vein. Sheetlike deposit of mineral matter within a fracture in rock.

Vesicle. Small cavity of irregular to spherical shape formed by a trapped gas bubble in lava; adjective: vesicular.

Volcanic ash. Unconsolidated, explosively fragmented volcanic material of particle diameter less than ⅛ inch.

Volcanic bomb. A glob of lava ejected while still plastic, shaped and cooled in flight.

Volcanic cinders. Glassy, porous fragments of lava explosively ejected from a volcanic vent, from pea to baseball size.

Volcanic rock. Glassy to finely crystallized lava, or fragmented lava debris, erupted from a volcanic vent.

Water gap. Narrow gorge across a mountain ridge in which a stream flows.

Water table. Top of the subsurface zone that is saturated with water.

Wave-cut platform. Gently sloping surface cut along shore by wave erosion.

Wavelength. Crest to crest distance between successive waves in a train.

Weathering. Chemical decomposition and mechanical disintegration of rocks and minerals through interaction with the atmosphere and biosphere.

Wind gap. Saddle in crest of a ridge, now abandoned by the stream that cut it.

Wind ripples. Asymmetrical current ripples in wind-deposited sand.

Sources of Supplementary Information

Elementary Physical Geology

Birkeland, Peter W., and Larson, Edwin E. *Putnam's Geology.* New York, Oxford: Oxford University Press, 1989. A well-written classic, updated by two younger geologists; fundamental and sound.

Robinson, Edwin S. *Basic Physical Geology.* New York: John Wiley and Sons, 1982. Good on plate tectonics; emphasizes basic principles and how things work.

Landforms and Surface Processes

Easterbrook, Donald J. *Surface Processes and Landforms.* New York: Macmillan Publishing Co., 1993. A first-class book on the geological processes, such as weathering, streams, wind, glaciers, tectonics, shorelines, and groundwater, active in creating the scenery of Earth's surface.

Ritter, Dale F. *Process Geomorphology.* Dubuque: Wm. C. Brown Co., 1978. A good companion to Easterbrook's volume, with a somewhat different emphasis.

Shelton, John S. *Geology Illustrated.* San Francisco: W. H. Freeman and Co., 1966. A truly striking series of low-altitude and oblique air photos of classical geological features in the United States, with text explanations.

Minerals and Rocks

Court, Arthur, and Campbell, Ian. *Minerals, Nature's Fabulous Jewels.* New York: Harry N. Abrams Inc., 1974. A collection of spectacular photographs of minerals with short, pithy, descriptive paragraphs.

Dietrich, Richard V., and Skinner, Brian J. *Rocks and Rock Minerals.* New York: John Wiley and Sons, 1979.

Fenton, Carroll, L, and Fenton, Mildred, A. *The Rock Book.* New York: Doubleday Doran and Co., Inc., 1940. Old but with still pertinent simple descriptions, not overly technical. Well illustrated.

Geology of California

Bailey, Edgar H., ed. 1966. *Geology of Northern California.* California Division of Mines and Geology, Bulletin 190, 1966. A compendium of articles by many authors on the northern half of the state.

Jahns, Richard H., ed. 1954. *Geology of Southern California.* California Division of Mines and Geology, Bulletin 170, 1954. A compendium of a host of articles by many authors touching on the geology of all natural provinces and specifically on important features therein.

Norris, Robert M., and Webb, Robert W. *Geology of California* (second edition). New York: John Wiley and Sons, 1990. The best book ever written about the regional geology of California, natural province by natural province. An excellent revision by author Norris.

United States Geological Survey publishes many books and maps on California geology. College libraries maintain collections.

Books can be purchased from:
U.S. Geological Survey, Book Sales
Box 25286
Denver, Colorado 80225

Maps from:
U.S. Geological Survey, Map Sales
Box 25286
Denver, Colorado 80225

Information on state publications and maps available from:

California Division of Mines and Geology
Publications and Information Office
801 K Street, 14th Floor, Mail Stop 14-32
Sacramento, CA 95814-3532
Publication and Maps of C.D.M.G. are available for over-the-counter sale at regional offices in San Francisco and Los Angeles.

California Geology, a bimonthly pamphlet published by the California Division of Mines and Geology.

Order from:
California Geology
P.O. Box 2980
Sacramento, CA 95812-2980
This publication provides reports on C.D.M.G. projects and articles on features of California geology of general public interest, plus news items on earth sciences in California.

Geological Map of California (small). A dandy, affordable one-shot generalized map of the state, in color, showing rock types and distribution, major faults, and natural provinces of the whole state.

Order from:
California Division of Mines and Geology
P.O. Box 2980
Sacramento, CA 95812-2980

Guidebooks

Alt, David D., and Hyndman, Donald, W. *Roadside Geology of Northern California.* Missoula, MT: Mountain Press Publishing Co., 1975. A succinct, well-illustrated guidebook to features, areas, and routes in the northern half of the state.

Sharp, Robert P. *Geology. A Field Guide To Southern California.* Dubuque: Kendall/Hunt Publishing Co. A detailed field guide to the natural provinces and some major routes of travel in the southern part of the state.

Smith, Genny S., ed. *Deepest Valley.* Mammoth Lakes, CA.: Genny Smith Books, 1978. Articles by multiple authors on natural and human history with information on roads and trails in Owens Valley. A classical assemblage of useful and interesting information.

Smith, Genny S., ed. *Mammoth Lakes Sierra.* Fifth edition. Mammoth Lakes, CA: Genny Smith Books, 1989. Provides roadside and trail guides for the greater Mammoth Lakes area plus natural and human history, by multiple authorities.

Many field guides on southern California geology have been published by a host of professional organizations. These guides are hard to come by because of limited editions and distribution. Some college libraries have established at least fractional collections.

Local geological societies publish road guides and occasionally compendiums of articles on specific areas. A good example of the latter is:

Geology and Mineral Wealth of the California Transverse Ranges (Mason Hill Volume), 1982, from the South Coast Geological Society, Inc., P.O. Box 10244, Santa Ana, CA, 92711.

Index